MANHATTAN TO MINISINK

Also by Robert S. Grumet

Native Americans of the Northwest Coast (1979)
Native American Place Names in New York City (1981)
The Lenapes (1989)
Historic Contact (1995)
Northeastern Indian Lives (ed.; 1996)
Journey on the Forbidden Path (ed.; 1999)
Bay, Plain, and Piedmont (2000)
Voices from the Delaware Big House Ceremony (ed.; 2001)
Revitalizations and Mazeways (ed.; 2003)
Modernity and Mind (ed.; 2004)
The Munsee Indians (2009)
First Manhattans (2010)

MANHATTAN TO MINISINK
American Indian Place Names in Greater New York and Vicinity

Robert S. Grumet

Munsee and Northern Unami Etymologies
by Raymond Whritenour

University of Oklahoma Press : Norman

Library of Congress Cataloging-in-Publication Data

Grumet, Robert Steven.
  Manhattan to Minisink : American Indian place names in greater New York and vicinity / Robert S. Grumet.
      pages cm
  Munsee and Northern Unami Etymologies by Raymond Whritenour.
  Includes bibliographical references.
    ISBN 978-0-8061-4336-1 (hardcover) ISBN 978-0-8061-6902-6 (paper)   1. Names, Geographical—New York (State)   2. New York (State)—History, Local.   3. Cities and towns—New York (State)   I. Title.
  F117.G78 2013
  974.7—dc23
                                                                  2012049859

The paper in this book meets the guidelines for permanence and durability of the Committee on Production Guidelines for Book Longevity of the Council on Library Resources, Inc. ∞

Copyright © 2013 by the University of Oklahoma Press, Norman, Publishing Division of the University. Paperback published 2021. Manufactured in the U.S.A.

All rights reserved. No part of this publication may be reproduced, stored in a retrieval system, or transmitted, in any form or by any means, electronic, mechanical, photocopying, recording, or otherwise—except as permitted under Section 107 or 108 of the United States Copyright Act—without the prior written permission of the University of Oklahoma Press. To request permission to reproduce selections from this book, write to Permissions, University of Oklahoma Press, 2800 Venture Drive, Norman OK 73069, or email rights.oupress@ou.edu.

For Bill

CONTENTS

List of Maps	viii
Preface	ix
Acknowledgments	xv
Timeline	xvii
Maps	xxi
Introduction	1
Munsees, Lenapes, Delawares, and Greater New York	3
Etymology and American Indian Place Names	7
The Influence of European Colonists on Place Names	9
What Are Indian Place Names Doing on Modern Regional Maps?	10
Problems of Translation and Identification	13
Structure and Organization	15
Part 1. Colonial-Era Indian Place Names	19
Part 2. Imports, Inventions, Invocations, and Impostors	191
Sources	247

MAPS

Following page xxii

The Munsee Homeland
New York City and Western Long Island
Mid-Hudson East
Mid-Hudson West
New Jersey and Pennsylvania
Passaic River Valley
Raritan-Navesink Country

PREFACE

I've learned a few unexpected things while working on this book. Up until the time my colleague and friend SUNY New Paltz history professor emeritus Laurence M. Hauptman suggested that I undertake this project, I rather smugly thought that my career-long devotion to collecting, organizing, and testing hypotheses explaining patterns abstracted from data documenting Munsee Indian history and culture in and around present-day Greater New York during colonial times had provided most of what I needed to finally write a definitive book on the region's Indian place names. I had, after all, compiled an extensively cross-indexed paper database built around chronologically arranged file folders holding texts of many thousands of transcribed and photo-reproduced documents, maps, and other data gathered over the years. Abstracts of information extracted from these data files were entered onto topically and temporally organized cross-indexed file cards. These, in turn, were used to create chronologically ordered file sheets tracing the documentary careers of particular people, places, and polities through colonial records from their initial appearances to their final disappearances.

This paper database was christened the Munsee File as a tribute to pioneering Canadian ethnographer C. Marius Barbeau's magnificent Tsimshian File (Cove 1985), presently curated by the Canadian Museum of Civilization in Gatineau, Quebec. Eager to take advantage of the opportunities for nearly instantaneous data query and retrieval offered by computers, I devoted the first years of my retirement to digitizing those parts of the Munsee File containing Indian deeds and related land records dating from 1626 to 1767. I knew from past experience that these records preserved a substantial number of the thousands of Indian place names mentioned in file documents. Relying on my computer to provide swift and efficient access both to this new body of computerized primary data and to the growing body of mostly secondary information available online, I expected work on this book to be a largely mechanical exercise that would go rather quickly.

As with so many other past smug assumptions, this one was promptly smashed. Reality set in with the first Munsee File data searches. The perfect storm that I encountered in earlier work of differently spelled names whose orthographies and associations often varied widely one from the other recurred again here. Once again, new efforts had to be made to link often widely divergent words written down by colonial scribes who followed no strict spelling conventions, whose tin ears often failed to fully pick up sounds even in their own languages, and who, when all was said and done, neither understood the Munsee language nor cared about its phonological or morphological rules.

Things did not become less confusing when I began examining what happened to the names after they were recorded. Most, and far more than expected, were not added onto colonial maps after colonial scribes wrote them down. They were, instead, simply filed away for safekeeping and were only brought up when land disputes put them into play. Sometimes Indian people were asked to confirm or deny associations linking one or another Native name to this or that place mentioned on a deed brought into evidence in a suit. Courts increasingly relied on non-Indian testimony as Munsee population dwindled in their homeland. As the research progressed it became clear that whoever employed Indian place names and however they were used, they had altogether independent lives, or, more accurately, afterlives, of their own that went on long after the people who uttered and recorded them were dead and gone.

Computer searches, web queries, and paper research that could, and did, often drag on for hours, days, sometimes weeks at a time shattered my belief that the work would go quickly as well as easily. Truth be told, I had already been disabused of this notion. Work I had started in college for my senior honors thesis on New York City Indian place names did not end with publication of a monograph based on these findings (Grumet 1981). Several place names eluded efforts to track down their origins and etymologies. One particularly refractory place name, Mosholu in my old neighborhood in the Bronx, only revealed itself when I stumbled over the entry for Mashulaville, Mississippi, in a recently published copy of Bright (2004:270) supplied as research began by University of Oklahoma Press acquisitions editor Alessandra Jacobi-Tamulevich.

Other assumptions also fell by the wayside as work continued on this book. Like almost everybody else trying to understand place names, I had always sought the holy grail of the one true translation. It had always been easy to breeze past the vast number of etymologies cobbled together by people who did not know anything about Indians or names. Only the smallest number of directly documented Indian place names, however, seemed to provide exact phonological matches for Indian words. Goddard (2010), for example, thinks Manahata and Manahatin, the earliest known orthographies of Manhattan, precisely reproduce the sound of the Munsee word *Man-ă-hă-tonh*, a name translated by a fluent Munsee-speaking traditionalist as "the place where timber is procured for bows and arrows." What, then, are we to make of Manados, Manahatesen, Manatus, Mannatens, and the host of other orthographies leading up to the present-day spelling that others, including several Indians, have translated as spellings closely resembling the Unami word *Mënating*, "island"? Many of these orthographies may, of course, be misspelled transmission errors of one sort or another. At least some, however, have a fair chance of being examples of universal patterns of play and drift that allow people to assign entirely different meanings to nearly identical-sounding words.

This capacity of names to embody multiple meanings poses challenges that researchers ignore at their peril. Original name meanings disappeared with their inventors. Subsequent users, both those descended from a name's originators and its foreign adopters, invest a name with meanings of their own that may or may not match those intended by its inventors. In many ways, the multivocal character of names is reflected in the variable spellings, unclear locations, and inconsistent translations recorded by all-too-fallible chroniclers. Linkages of similar-sounding words (such as Manahata and Manahatin) and explicitly recognized different words (thus and such, called by the Indians this or that) need to be cross-checked and verified by date, location, and historical context. Patterns formed through these linkages must be considered provisional and need to be tested whenever possible so that useful linkages can be validated and erroneous ones eliminated.

A last fondly held and deeply ingrained assumption finally fell away only during the most intensive phase of research into the documentary careers of the names in this book. I had always, it seems, rather unthinkingly regarded Indian place names as somehow more genuine toponyms having greater rights to their places on maps than those that replaced them. I knew, of course, that more than a few of the places presently bearing Munsee names had been identified by other, mostly European, names sometime in the past. I also knew that some of the names on present-day Greater New York maps were recent reintroductions of Indian names, some locally derived and some not. I thought that many early European successor names, especially the comic-sounding ones such as Grin Town and colorless, lackluster, more prosaic names such as Smith's Mills, were temporary aberrations corrected by restoration of original Indian names. I discovered, to my surprise, that a far larger percentage of recorded Indian place names had disappeared, and even more had been reintroduced to replace European names than I had expected. Discovery of these phenomena brought home the fact that Indian place names, like all cultural products, were malleable artifacts as manipulable as any others by politicians, developers, preservationists, scientists, community activists and action groups, local experts, and romantic dreamers.

Anthropologists Richard Handler and Jocelyn Linnekin (1984) have pointed out that rather than being either genuine or spurious (as Sapir [1949] suggested for culture), cultural traditions combine fact and fiction to serve present needs. You will thus not encounter genuinely true or spuriously false Indian place names or etymologies in this book. Instead, you will encounter two types of names characterized by their relative positions in time and space in Greater New York. The first section presents Indian place names chronicled in records created during colonial times. The second section describes those created later or imported onto the region's maps from other places.

This approach should more clearly show how much the region's toponymy has changed over the past four centuries since Europeans first starting putting place names on their maps. It will also help introduce some of the people who have placed and replaced Indian names on these maps, figures such as Victor Guinzburg, the turn-of-the-century jokester who invented the name Lake Heaptauqua; local historian William N. Nelson, whose self-discovered erroneous Dutch etymology Naachpunkt of the Indian name of the sachem Nachpunk made a century ago is still on modern-day maps; and inventive types such as Henry Rowe Schoolcraft and his Ojibwa wife, who filled empty spaces across New York with Ojibwa names that they thought the region's first people might have sanctioned. You will also encounter mythmongers such as Robert and Reginald Pelham Bolton, and the helpful folks at the Fairfield Historical Society, who in 1952 worked with a local developer to place the largest single group of Indian names onto one locale in the region. These and the other toponymic resurrectionists who have found places in this book's pages represent only a tantalizingly small fraction of the many place name movers and shakers whose stories have yet to be told.

In the end, findings presented in these pages are less a definitive last word and more like an interim report on the current status of knowledge of one class of place names in the Greater New York area. Much more detailed investigations into the region's Indian names and their philologists, such as the nineteenth-century Canadian Munsee scholar Albert Anthony, who championed the aforementioned *Man-ă-hă-tonh* etymology for Manhattan, await the attentions of future researchers. I am also sure that, just as present-day archaeologists carefully sift through refuse that their predecessors discarded as garbage, future investigators will carefully troll through naive and folk etymologies currently considered spurious constructs, as well as what are regarded as more generally accepted genuine Native products in search of insights into the minds and intentions of place name creators and users. By so doing, our successors will, I am sure, find themselves unable to avoid changing much, if not everything, we think we know about names today.

Chances for new discoveries increase every day as information originally obtained orally and transcribed onto paper proliferates electronically. Right now, people regarding Indians as embodiments of their fondest hopes and dreams are searching out old Indian names and putting them on maps in record numbers. Because of this, we are living at a time that future generations may remember as a golden age of local Indian place name restoration. There is no telling how long this trend may last—perhaps until all of the thousands of Indian names still lying unacknowledged on old manuscript pages are restored to area maps. Perhaps place names' chameleon-like ability to carry the weight of unintended meanings and unexpected significance may result in the transformations of even the most everyday examples into solemnly venerated vessels of tradition used to sustain claims for reparations for suddenly sacred lands. Whatever happens, the liminally fraught nature of names all but ensures that they will be swords that cut all ways. Who, then, can say what place names, Indian and otherwise, will mean to future generations in a world where change, as the cliché goes, is the only constant?

ACKNOWLEDGMENTS

I am grateful to colleagues and friends who have assisted me in this project. Laurence M. Hauptman, who first suggested that I write this book, and John Bierhorst, Scott M. Chapin, Mary C. Davis, Joe Diamond, Henry L. Gitner, Ives Goddard, Lion Miles, James S. Muncey, Stuart A. Reeve, and James Rementer have done their best to detect errors and provide useful comments. Bruce L. Pearson gave the final draft of the manuscript a very close reading; his efforts significantly improved the final product and are appreciatively acknowledged here. Not being a linguist, I have continued to rely on the generosity of colleagues specializing in the study of the Delaware, German, and Dutch languages essential to reconstructing Indian culture and history in this region. New Netherland Project director Charles T. Gehring, who has cheerfully rendered indispensable assistance in deciphering early modern Dutch language materials for many years, has again helped out here. Munsee and Northern Unami philologist Raymond Whritenour has generously supplied etymologies and provided much constructive criticism. And once again, I thank David K. Richter and the staff of the McNeil Center for Early American Studies at the University of Pennsylvania for the Senior Research Associateship that makes scholarly resources available to a retired and otherwise unaffiliated researcher.

 The staff at the University of Oklahoma Press, particularly acquisitions editor Alessandra Jacobi-Tamulevich and managing editor Steven Baker, have done their usual fine job of keeping me on the straight and narrow. Mapping Specialists Ltd. project manager Jeff King oversaw development of the volume's maps. Copyeditor Lori Rider patiently ferreted out anomalies and recommended remedies to correct the many examples of what one of my graduate school advisors rather delicately described as editorial infelicities that managed to elude both my eyes and the scrutiny of those who looked over the final draft. All those that survive remain mine alone, and I bear full responsibility for them.

TIMELINE

1609   A seaman serving under Henry Hudson, an Englishman in Dutch service, records the first Indian place name in Greater New York. It is Manahata, today's Manhattan Island.

1626   The Dutch West India Company's New Netherland colony governor, Minuit, buys Manhattan from the Indians for sixty guilders' worth of trade goods. He builds Fort Amsterdam at the island's southern tip; the village of New Amsterdam soon rises around the fort.

1637   The Pequot War in nearby Connecticut devastates the Pequot Indian nation.

1640   The conflict known today as Governor Kieft's War breaks out between settlers and Indians along the Lower Hudson River. Devastated by colonial attacks, the Lower Hudson River Indians make peace with the Dutch in 1645.

1655   The Peach War begins in New Netherland. Hostilities embroiling settlers and Indians drag on until 1657.

1659   Fighting breaks out between settlers and Indians at Esopus. The Esopus Wars finally end in 1664.

1664   The English conquer New Netherland. The provinces of New York and New Jersey are established on the former Dutch lands.

1665   New York's Governor Nicolls negotiates a treaty of friendship with the Esopus Indians. Nicolls's treaty terms require annual renewal.

1672   The Third Anglo-Dutch Naval War begins. The Dutch seize New York and New Jersey. The treaty ending the war in 1674 restores the colonies to English rule.

1681   Pennsylvania is established as a proprietary colony granted to William Penn.

1689   The First French and Indian War, also known as King William's War, begins. Fighting stops in 1697.
1692   Shawnees in the Mississippi Valley who have fought alongside the French change sides and accept a Minisink Indian invitation to resettle at the Delaware Water Gap. They move by 1694.
1702   The Second French and Indian War, also known as Queen Anne's War, breaks out. Fighting stops in 1713.
1727   The Munsees are first mentioned by name in colonial records. The New Jersey Indian king Weequehela is hanged for murdering a settler. Most of his people move to the Forks of the Delaware.
1728   The Shawnees abandon the Delaware Water Gap and move west.
1737   The Walking Purchase allows the Penn family to claim and seize Indian lands at the Forks of the Delaware.
1739   Protestant refugees belonging to the Moravian Brethren move to the Forks and begin building their main settlements of Bethlehem and Nazareth. They soon start Indian missions at Shekomeko in Dutchess County, New York, and Pachgatgoch at Schaghticoke, Connecticut.
1744   The Third French and Indian War, also called King George's War, begins. David Brainerd begins his mission among the Forks Indians.
1745   Brainerd's Munsee and Northern Unami followers move back to New Jersey.
1746   Dutchess County authorities force Moravians and their Indian converts at Shekomeko to move to the Forks.
1748   The Third French and Indian War ends.
1754   The Fourth French and Indian War, also known as the Seven Years' War, begins. Munsees and other Delawares launch attacks against the New York, New Jersey, and Pennsylvania frontiers. Protracted conflict known as the Yankee-Pennamite Wars begins when Connecticut colonists purchase northeastern Pennsylvania from the Indians in deeds signed in 1754 and 1755.
1758   Treaties signed at Easton, Pennsylvania, restore peace between the British and most Munsees and their Indian allies. Indian land claims in New Jersey are adjudicated, and a tract set aside for them named the Brotherton Reservation is established at Edgepillock in present-day Indian Mills. France and Great Britain make their own peace in 1762.
1775   The Revolutionary War breaks out between the colonies and Great Britain. Although the war formally ends with American independence in 1783, continuing frontier violence forces Munsees to give

	up attempts to reestablish themselves on ancestral lands by the beginning of the nineteenth century.
1801	The Brotherton Reservation is dissolved. Most of its residents join their brethren at the Oneida Reservation in upstate New York. Many ultimately wind up in Wisconsin, Kansas, or Ontario.
1812	The War of 1812 with Great Britain begins. Fighting ends in 1815.
1828	Rediscovered Indian names begin to travel when the canal era begins in the region. Work on the Delaware and Hudson Canal begins (operations cease in 1899). One year later, work starts on the Lehigh Canal (the canal closes in 1940). Construction on the Morris Canal begins in 1831 (canal navigation stops in 1924). Work on the Delaware and Raritan Canal starts in 1834. The canal closes in 1931.
1832	The railroad era in the region begins with the organization of the New York and Erie Railroad. Railroads can carry Indian names farther and faster. Reconstituted as the Erie Railroad in 1861, the line merges with the Delaware, Lackawanna, and Western Railroad to form the Erie Lackawanna line in 1960. Acquired by Conrail in 1976, portions of the line's passenger service acquired by New Jersey Transit in 1983 continue to operate.
1834	The Long Island Railroad is established.
1846	The Pennsylvania Railroad and the Lehigh Valley Railroad are authorized. The Lehigh main line is completed by 1855. The Penn Central Railroad acquires most route miles of both lines in 1968, and Penn Central is absorbed by Conrail in 1976. Portions are currently administered by the CSX and Norfolk Southern corporations.
1849	The Central Railroad of New Jersey is established. The line is absorbed into Conrail in 1976. Work begins on the Belvidere-Delaware Railroad line in 1851. Acquired by Conrail in 1976, remnants of trackage are currently operated by CSX and a number of local short lines.
1858	Hackensack and New York Railroad operations begin. The line merges with the Erie Railroad in 1896 and currently operates as New Jersey Transit's Pascack Valley Line.
1861	The Civil War begins. Fighting ends in 1865.
1866	The Ulster and Delaware Railroad is founded. Acquired by the Penn Central Railroad in 1968, most of the line taken over by Conrail in 1976 is abandoned by 1980. The New York Central Railroad is formed from existing lines in 1867. Its West Shore Railroad division is established in 1885. In 1968 the New York Central merges with the Pennsylvania Railroad to form the Penn Central Railroad. Most

of the line's trackage acquired by Conrail in 1976 is operated today by CSX and Norfolk Southern.

1880 The New York, Ontario, and Western Railroad starts service. Operations cease in 1957. The Sussex Valley Railroad (established in 1867) and several other lines merge in 1881 to form the New York, Susquehanna, and Western Railroad. In 1999, parts of the line still in service are mostly divided between the CSX and Norfolk Southern lines.

1898 Manhattan, Brooklyn, Staten Island, parts of Westchester (present-day Bronx borough), and Queens are consolidated into the City of New York.

1956 Congress passes the Interstate Highway Act.

1971 Amtrak is formed as a federally owned government corporation administering intercity rail service.

1976 Conrail is established to federally administer bankrupt rail lines. The last Conrail lines are returned to private ownership in 1999.

1983 New Jersey Transit (formed in 1979) begins operating Conrail passenger service lines in the region.

MAPS

GENERAL MAP KEY

Indian names dating to the region's colonial times appear in bold-faced capital letters. Imported, invented, and non-Indian names appear in gray type.

The Munsee homeland

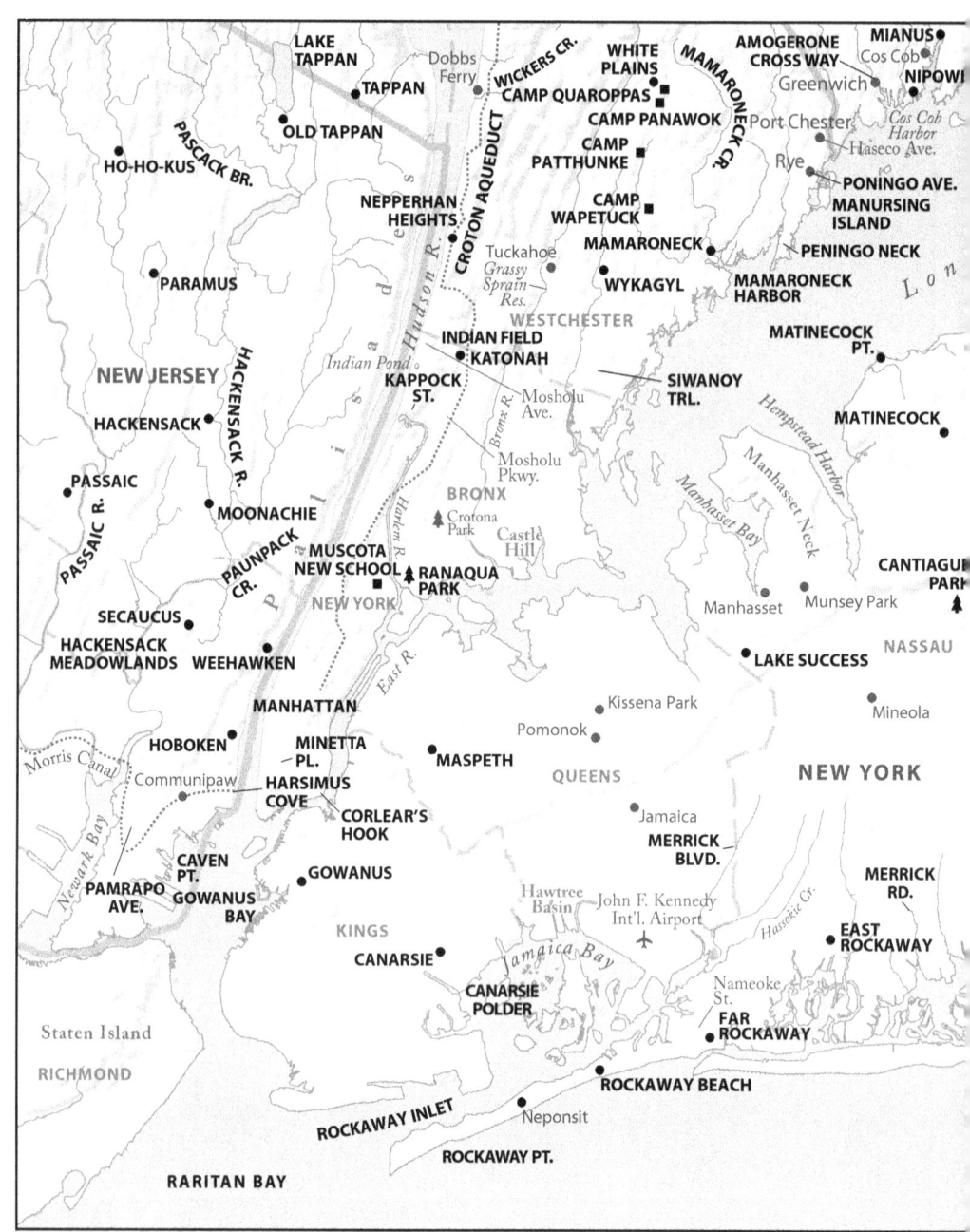

New York City and Western Long Island

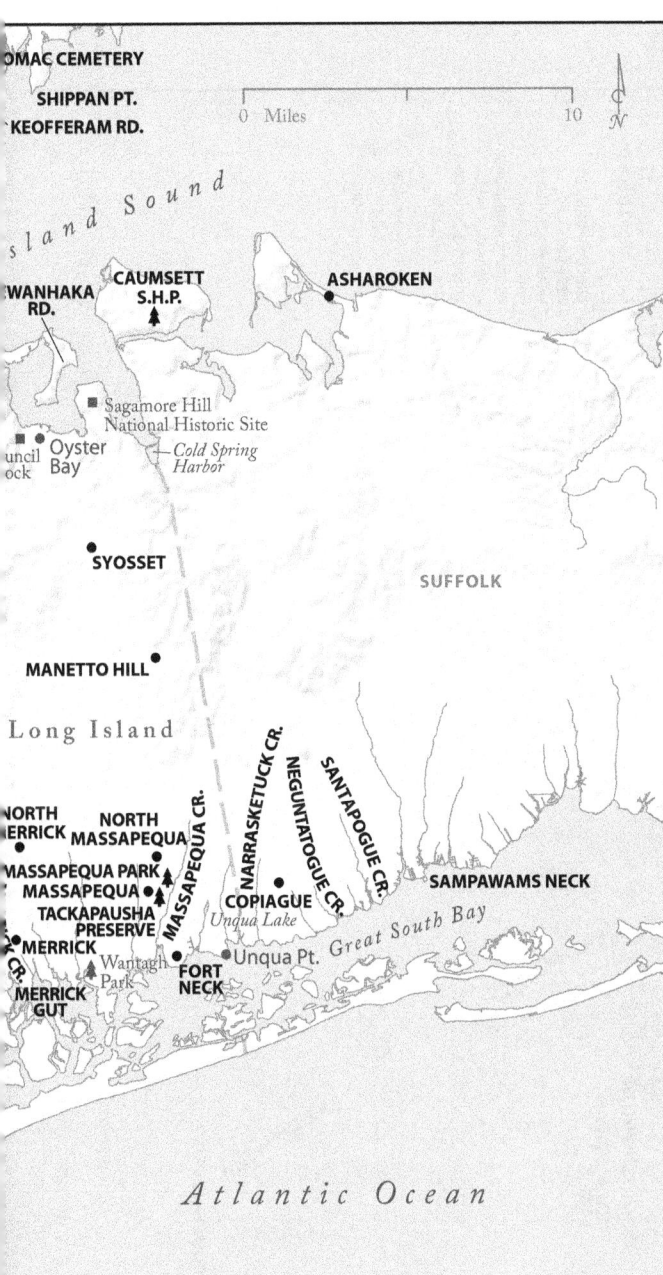

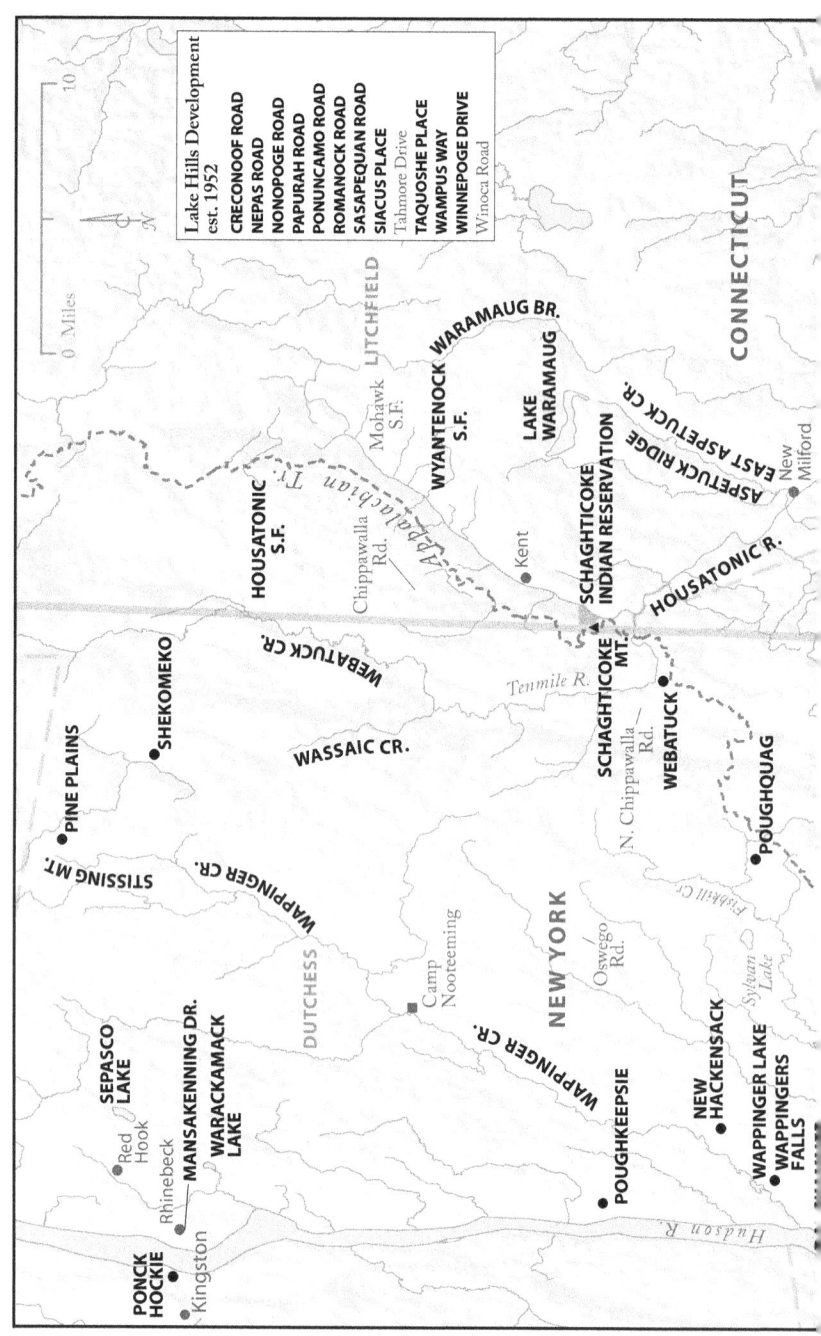

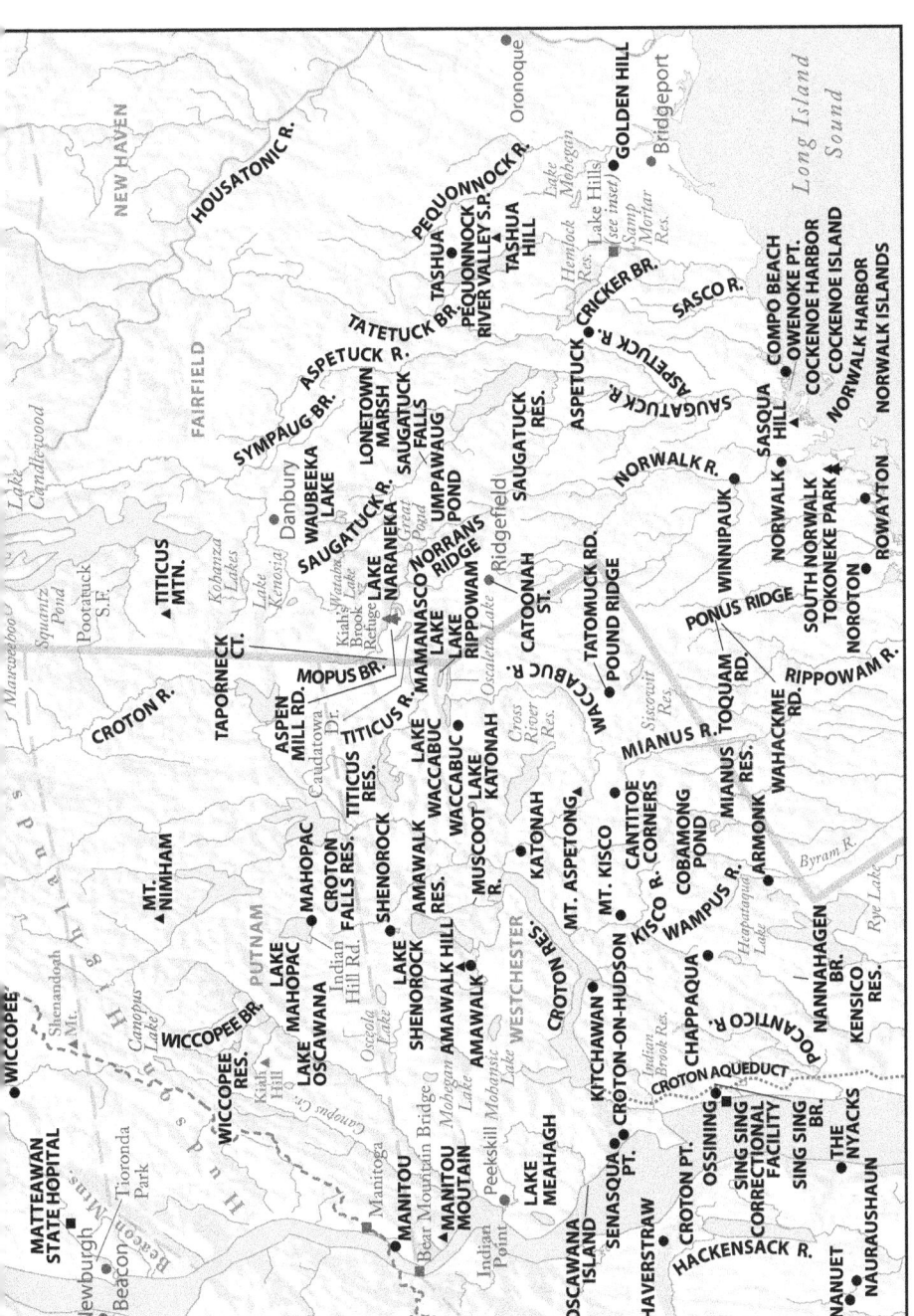

Mid-Hudson East

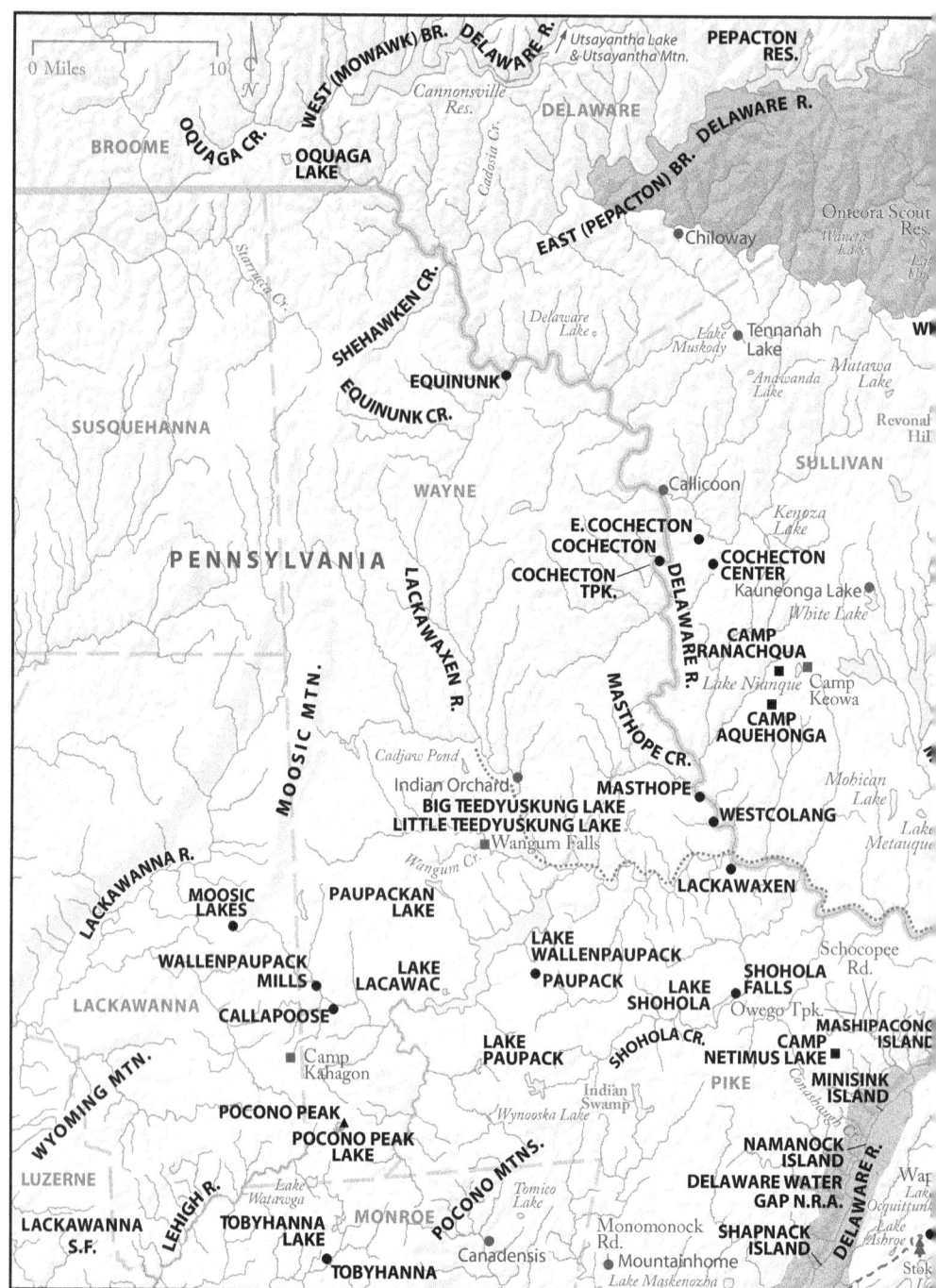

Mid-Hudson West

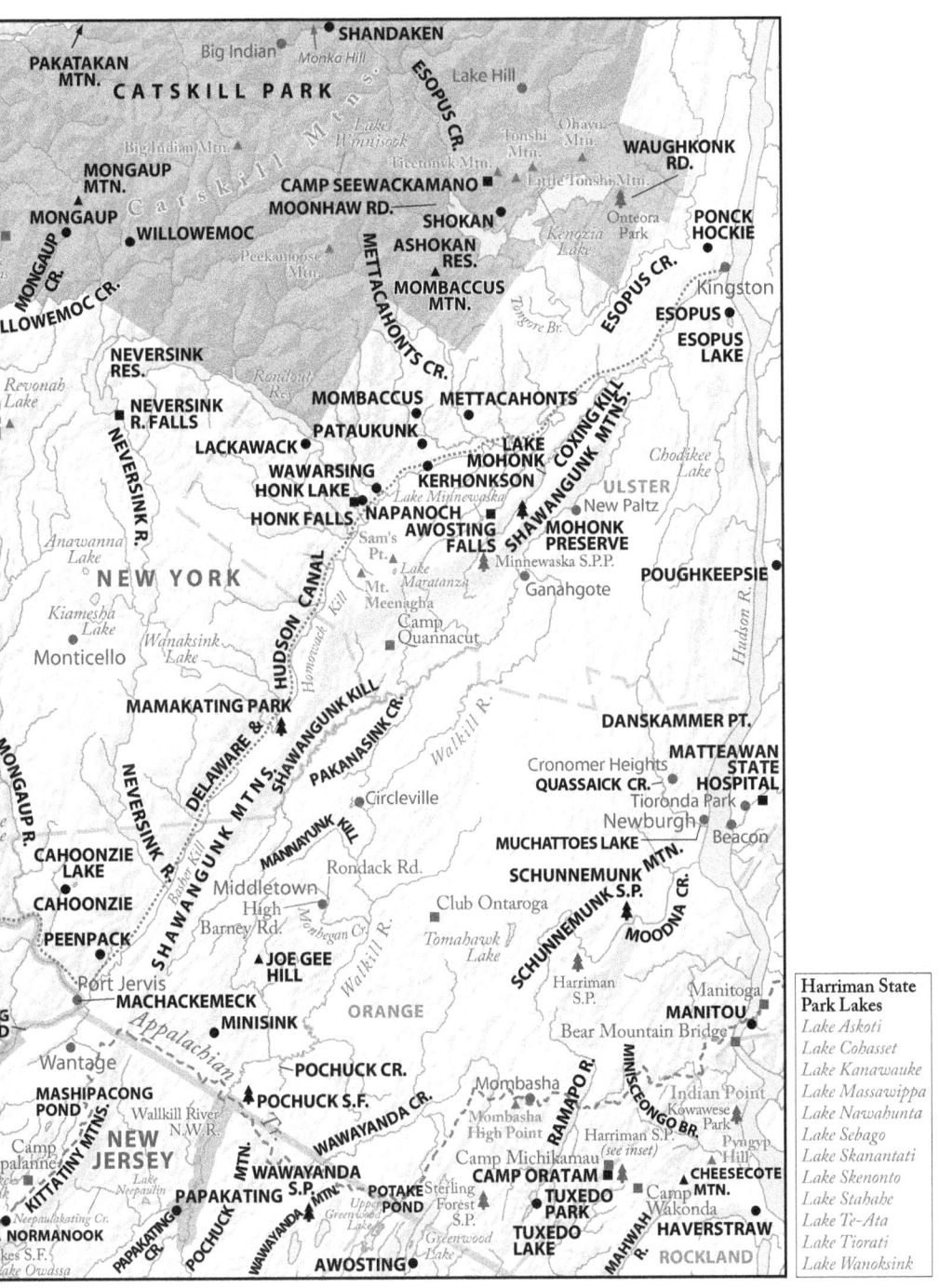

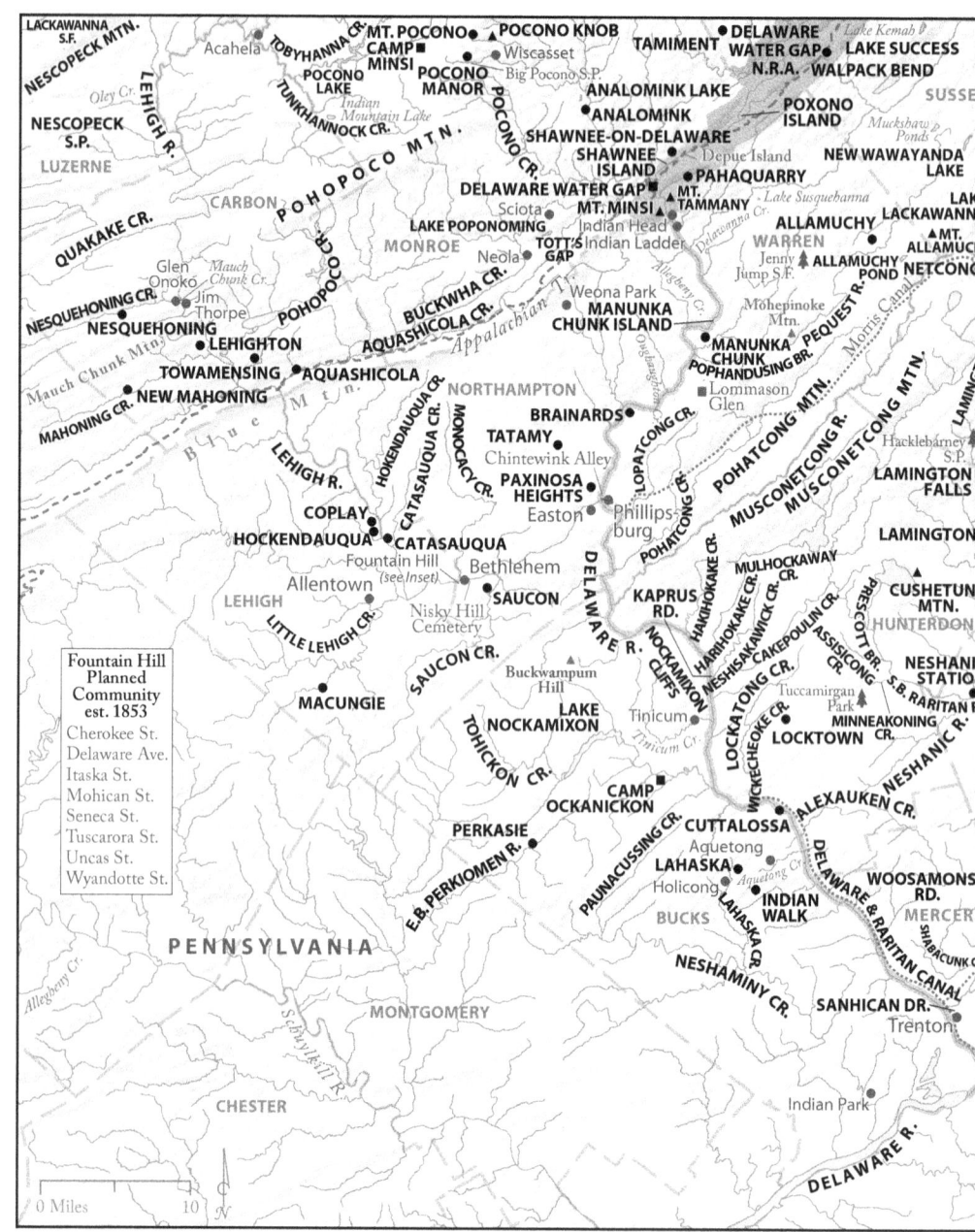

New Jersey and Pennsylvania

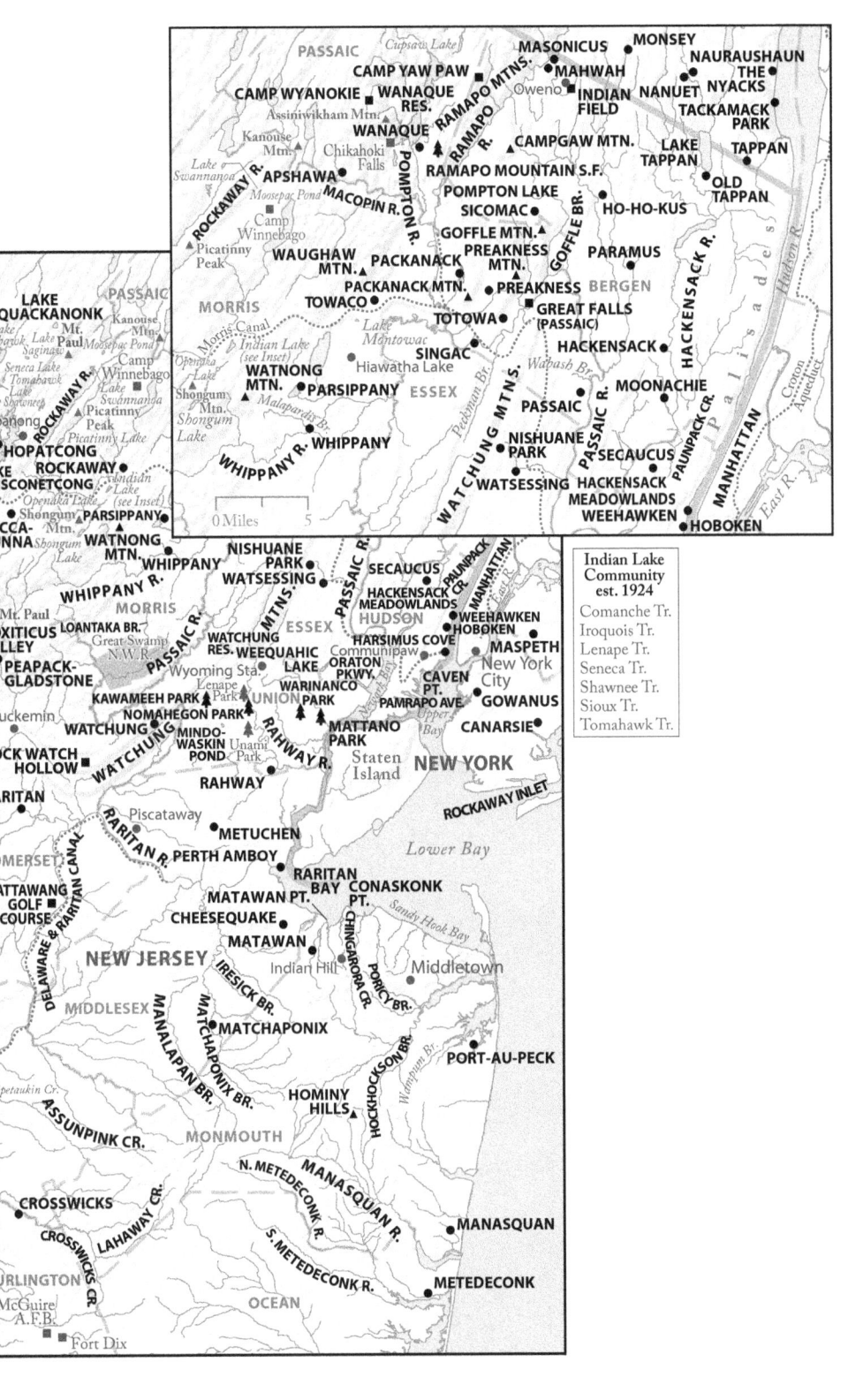

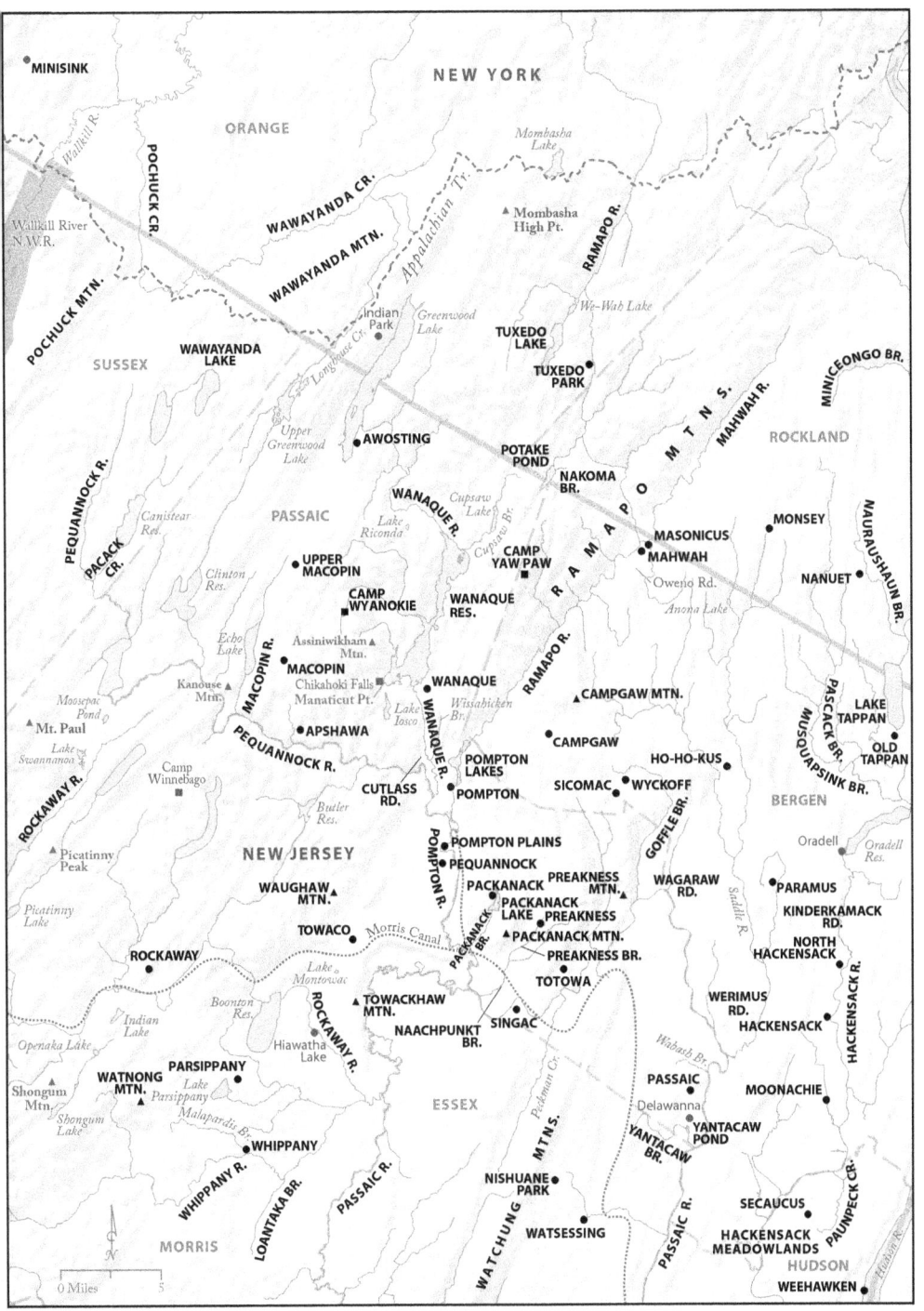

Passaic River Valley

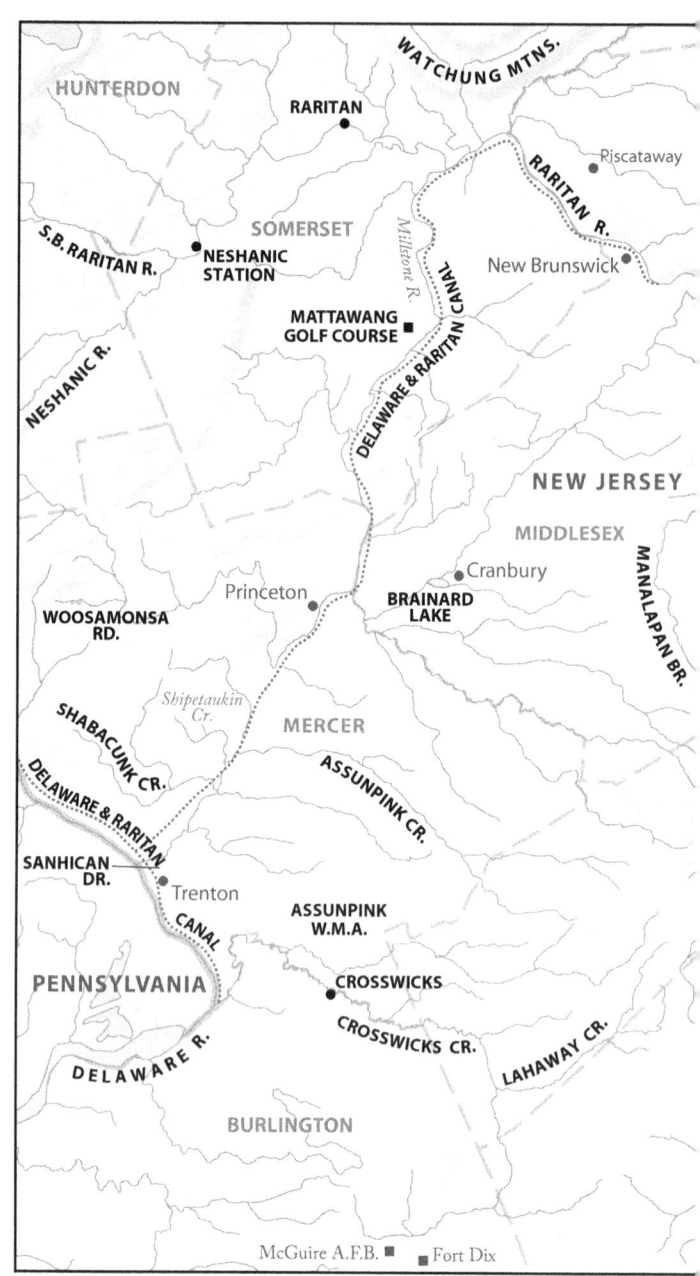

Raritan-Navesink Country

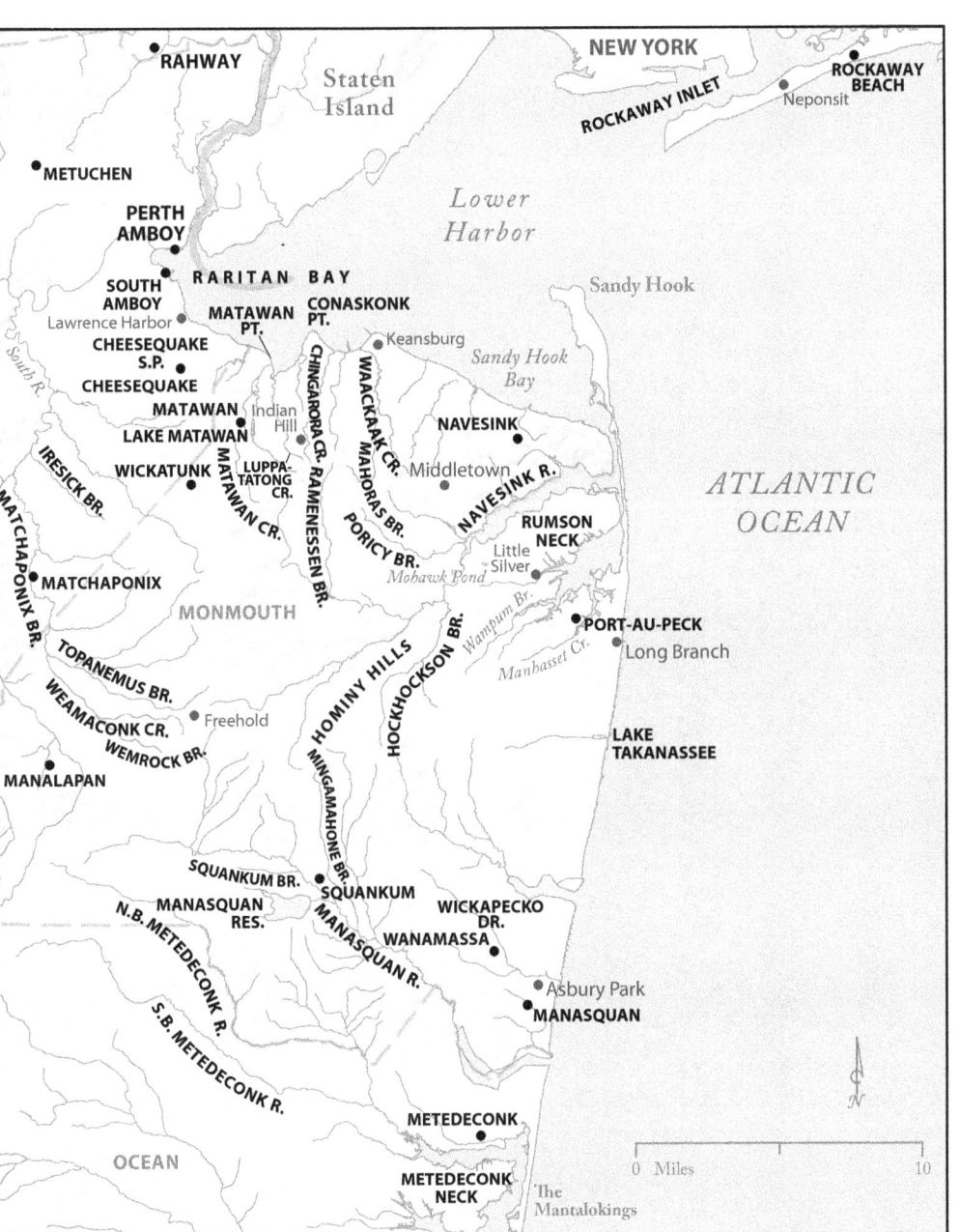

INTRODUCTION

This book is a guide to the more than five hundred actual and invented Indian place names currently on Greater New York maps. The area covered by this survey lies between the mountains and the sea along a broad swath of Middle Atlantic seaboard that stretches from Manhattan Island and its lower Hudson and western Long Island hinterland, across the Highland and Piedmont valleys of northern New Jersey and southeastern New York, to the Minisink country and the Lehigh Valley where the Upper Delaware River flows between northwestern New Jersey and northeastern Pennsylvania. This region closely matches the boundaries of the historic homeland of the Munsee Indians, the first people of today's metropolitan area.

Geography has set its physical limits on the bounds of this region ever since people first began coming to the lands between the Hudson and Delaware valleys more than eleven thousand years ago. The area's topography channeled human occupation between the rivers into the shallow basin drained by streams flowing from the Appalachian mountain ranges today called the Catskills, Poconos, and Berkshires. Pierced at strategic points by steep, narrow passes such as Storm King on the Hudson and the Delaware Water Gap farther west, these lines of hills run roughly parallel to the Atlantic Coast less than one hundred miles away. They are not the only physical barriers standing guard over the region. Sandy pine barrens in central parts of Long Island and New Jersey take over where the mountains recede from the coast at the region's eastern and southernmost limits.

The Indians who encountered European voyagers sailing to the region during the 1500s and early 1600s followed what archaeologists call a Late Woodland way of life. Late Woodland people lived in village-centered family-based societies supported by forms of farming, fishing, foraging, and hunting that had first emerged in the region about a thousand years earlier. Like other Woodland people everywhere across the Northeast, Munsees set up their mat- and bark-covered houses framed with lashed bent saplings in clearings alongside one or another of the many rivers and streams that coursed through their homeland. Travelers paddling their wooden dugout canoes along these waterways forged and maintained kinship and marriage ties that linked networks of farming towns, fishing camps, and hunting lodges. The configurations made by these networks revealed on maps reconstructing regional settlement patterns most nearly resemble pearly strings of tiny island archipelagoes cut out of the sea-green expanses of an otherwise all but unbroken woodland forest canopy.

The colonial newcomers would have readily recognized the Late Woodland lifestyle. Many had already encountered people following similar lifeways elsewhere in the world. Families of voyagers from more rural communities still lived in thatch-roofed, open-hearthed cottages not at all dissimilar to the mat- and bark-covered dwellings of the Indians they visited. More than a few could remember their own immediate ancestors' reliance on tools and weapons used by Indians, such as the bow and arrow, which was still employed with lethal success by Ottoman bowmen as recently as the Battle of Lepanto in 1571. Archaeologists recently recovered 137 longbows and more than 3,500 arrows from the wreck of the *Mary Rose,* a warship that sank in Portsmouth Harbor in 1545.

Then as now, people following different lifeways lived beyond the boundaries of the region. During Late Woodland times, Mohawks and their Iroquois confederates, speaking Iroquoian languages as different from Delaware as English is from Japanese, lived north and west of the Catskills. When the Europeans first arrived, they found the Iroquois to be a dynamically expansive people organized into a confederation of five nations (a sixth, the Tuscaroras, joined by 1722) that were among the first Indians in the region to acquire firearms. Algonquian-speaking neighbors who suffered from their attacks came to call them Iroquois ("little snakes" in its Ojibwa or Ottawa forms) and the easternmost of their nations as Mohawks (an Algonquian conjugate of the Southern Unami word *mhuweyok,* "cannibal monsters"; they called themselves *Kanyékeha:ka,* "people of the place of the flint").

Farther east, communities of Mahicans (many of whose descendants today call themselves Mohicans and who should not be confused with the altogether different Mohegans of eastern Connecticut) and other people speaking Eastern Algonquian languages similar to Delaware clustered at intervals along the northeastern borders of the Munsee homeland. They formed a vast arc stretching from the far slopes of the eastern Catskills, across the Upper Hudson

Valley and over the Berkshire Mountains to the Housatonic Valley, thence south to eastern Long Island. Beyond them lived the Pequots, Narragansetts, and other populous and powerful southern New England Algonquian Indian nations. Farther south, Delawares speaking closely related Unami dialects lived alongside streams flowing through the Jersey Pinelands and the Lower Delaware Valley. Beyond them lived Algonquian-speaking Nanticokes, Conoys, and Powhattans.

Indians in Munsee country found themselves increasingly pressed by foreigners as the seventeenth century began. Newcomers in the employ of merchants from Holland built their main fort at the southern tip of Manhattan Island after buying the island in 1626. Naming the fort and the settlement that grew around it New Amsterdam, they almost immediately began trying to spread outward. There they collided with English settlers—expansion-minded New Englanders who established their first colonies during the first decades of the seventeenth century farther north, and Virginians who had built their own settlements around the same time to the south. Rivalries began when they butted up against one another along these frontiers. Settlers from Connecticut and Pennsylvania, for example, claimed the same broad swath of the present-day northern part of Pennsylvania stretching west above the Delaware Water Gap as included in the royal charters granted to both colonies. Contending colonists defended their claims to the region in a struggle now called the Pennamite-Yankee Wars. These got their name for the Pennsylvanian proprietary, Quaker, royalist, and German interest groups who put their differences aside to deal amicably with one another to form a united front countering the common threat posed by the Yankee New Englanders. The wars began soon after New Englanders purchased Indian lands above the Water Gap in 1755 from Delaware signatories who identified themselves as Ninneepauues. Sputtering on fitfully for more than four decades, the conflict ended only in 1799 when the New Englanders finally accepted Revolutionary War military reparation lands in northern Ohio's Western Reserve and Firelands districts in exchange for lands in Pennsylvania they had formerly claimed.

Munsees, Lenapes, Delawares, and Greater New York

More than 250 years have passed since Indian people last used words in their own languages to talk about places and names in their Greater New York homeland. You should not be confused if you do not recognize the name Munsee, which for much of the past two centuries was only used by a few people. Most of these were exiles living farther west whose ancestors were originally from the Greater New York area. The rest were the few linguists, ethnologists, archaeologists, and historians interested in their language, history, and culture. No document preserves evidence that Indians used the word Munsee in Greater New York while they still lived in their homeland. Colonial authorities first formally noticed the name when reports that "Munscoes" Indians living along the upper branches of the Susquehanna had murdered traders there filtered into Philadelphia in 1727. Like many, but not all, charges laid at the door of Munsee people during the violent years that followed, this one turned out to be false (the traders had been attacked but not killed).

The bad feelings that fueled such rumors were real enough, however. The name Munsee, which most modern-day linguists think means "people from Minisink" (Munsee for "at the island"), makes a pointed reference to one of the last refuges that the region's first people stubbornly clung to in their homeland. It took more than a century and a half of deed agreements fair and foul to wrench their hold from their lands. Embittered Munsees joining vengeful war parties that ravaged the region for much of the last half of the eighteenth century made sure that Minisink also became a colonial byword for sorrow and loss. The violence of this struggle generated an enduring sense of resentment that eased only as the generations that lived through its horrors passed on.

The name Minisink itself was common enough. Similarly named places in the Munsee homeland include the long-forgotten and decidedly uninsular "land of Minnissingh" in present-day Poughkeepsie; the town of Minisink in Orange County, New York; Manursing Island, buried under landfill beneath Westchester's Rye Playland; and perhaps even Manhattan, though that last one is still the subject of some debate. The Minisink of the *Munsiiwak* (a philological plural equivalent of Munsee), however, was and still is the name of their much-cherished island and surrounding valley nestled where the Upper Delaware River flows between New Jersey's Kittatiny Ridge and the Pennsylvania Poconos some ninety miles west of Manhattan.

Today, people mostly use Munsee in one of three ways. Its widest use is as a name of cities, towns, and neighborhoods, such as Muncy, Pennsylvania, and Muncie, Indiana, which stand like signposts along the many trails Greater New York's first people followed into exile to the Six Nations, Muncey Town, and Moraviantown reserves in Ontario, the Stockbridge-Munsee Reservation in Wisconsin, and the more distant reservations on the Great Plains in Kansas and Oklahoma. Anthropologists, historians, and some people who trace descent to Munsee ancestors often use it as a national name. Linguists employ the term when talking about the dialect still spoken by a few elders descended from these ancestors. Most classify it as the most northerly of several closely related tongues (Northern and Southern Unami are recognized today; Wampano farther east and Unalachtigo to the south may be others) that make up a language they call Delaware. Delaware was originally spoken by all Native people living between Delaware Bay and the Hudson Valley.

The homeland of the people who spoke the Munsee dialect of the Delaware language stretched across the sizable stretch of Mid-Atlantic seaboard mostly, but not entirely, bracketed between Manhattan and Minisink islands. Although we do not precisely know what they called themselves or their homeland, they probably used some of the words recorded by colonists such as the Moravian missionaries John Heckewelder and David Zeisberger. Both men had solid working knowledge of the Delaware dialects spoken by the Indians they worked among, first in and around the western reaches of the old Munsee homeland during the third quarter of the eighteenth century, and later in their places of exile in the Ohio Valley and beyond. Among the words they recorded were *Lenape*, "Indian or Delaware person," and *gemenhatteyummenna*, "our island," the Delaware word for what we call the continent of North America.

Delaware, the name of their language and the river draining the western portion of the Munsee homeland, is not an Indian word. Virginians sailing into Delaware Bay in 1610 named it and the river that flowed into it for their governor, Sir Thomas West, Baron de la Warr. Indian and European naming customs soon conspired to fix Delaware securely on maps of the river and the region. Both peoples tended to refer to themselves as residents of a particular place—a river, for instance, or a valley or a village. In many ways their names were the linguistic equivalents of seeds; those planted in favorable soil and under the right conditions could grow and branch out to cover wider areas.

Europeans preserved some names after they fell from use in the written records that tell us most of what we know about Indians. The Indians themselves, who almost totally relied on oral traditions for their own records, tended not to utter the real names of dead people (it was okay to mention their titles and nicknames). Names of many lost or abandoned places were also evidently not passed on. Those that stayed on maps, such as Manhattan, came to be so thoroughly associated with Europeans that quite possibly few considered them Indian names at all. Those confining their interests to living people and the places where they lived may simply have preferred to put the past behind them. Many may have chosen to forget irretrievably lost lands, friends, and relatives. Others perhaps worried that spirits hearing their names called might return to persuade the living to join them in the land of the dead that most Munsees thought lay beyond the edge of the western horizon.

Colonial documents in the Munsee homeland are all but silent on the subject of the word Lenape, the name widely used today by people when talking about Delaware Indians. It seems certain that most ancestors of the Munsees used the word *lunaapeew*, similar to the Northern Unami form *lennape* and *lënape*, the Southern Unami variant, when referring to themselves in their native language. A May 6, 1755, reference to Delaware Indians selling lands above the Delaware Water Gap as "the ancient tribe of Indians called Ninneepauues" (Boyd and Taylor 1930–71[1]:260–71) indicates that at least someone involved with Munsees living in the more northern parts of their homeland also utilized the Mahican N-dialect *anenapawa* (with the English pluralizer -s tacked on), equivalent to the Munsee *lunaapeew*, when talking to Europeans (in Masthay 1991:86). Lenapehoking, a name sometimes used today (it means "in the land of the Lenape"), came into use much more recently. Nora Thompson Dean, whose native language was a late nineteenth-century form of Southern Unami (which may have differed from the form first documented more than 150 years earlier much as modern American English differs from its colonial and old-country forebears), provided the name in response to a request made in 1984 by archaeologist Herbert C. Kraft (2001:9nn.) for a word resembling something her ancestors may have used for their homeland. Dean's ancestors were doubtless among Delawares who gave Heckewelder several other generic names for their homeland. These included *Scheyichbi*, "along the water's edge or seashore," used when talking about New Jersey (Heckewelder 1876:51), and *Laaphawachking*, "place of stringing beads," referring to coastal New York (in Robert Bolton 1881[1]:2).

The tradition of supplying substitutes for unrecorded Munsee place names is an old one in Greater New York. One of the first to do it was nineteenth-century Upper Great Lakes Indian agent and pioneering ethnologist Henry Rowe Schoolcraft (1845:22–27). Schoolcraft's first wife was a Native woman fluent in her people's Central Algonquian Ojibwa language. Working with Ojibwa words she supplied or verified, Schoolcraft grafted what they imagined might be her people's equivalents of long-lost Munsee names onto local maps. He renamed the Hudson River *Shatemuc*, "pelican river," rechristened Ellis Island *Kioshk*, "gull island," and gave the name *Ihepetonga* ("high sandy banks") to Brooklyn Heights. Popular for many years, these names and others the couple invented have been abandoned or, in a few cases, replaced by more demonstrably authentic local Indian names.

The Schoolcrafts did not create Ojibwa equivalents for otherwise unrecorded Munsee names for Long Island Sound and the Atlantic Ocean, but this did not prevent others from looking for their Indian names. Another pioneering ethnologist, Lewis Henry Morgan (1851:474), found out that

Iroquois people called the Atlantic *Ojikhadagega,* "salt water." The Munsee word *ketauhekan,* "the ocean or sea," occurs as a common noun in a Canadian Munsee prayer book first published in 1842 (Wampum 1886).

What we know about other names that settlers and Indians used to identify the two river valleys lying at the opposite sides of the Munsee homeland reveals an only slightly less derivative namescape. Unami people speaking downriver Delaware dialects called the Delaware River *Lenapewihittuck,* "River of the Lenapes." It was *Makerick Kitton* in the pidgin Delaware trade jargon spoken throughout the region during the 1600s that Heckewelder thought meant "strong rapid or a rapid stream"; Munsee language specialist Ray Whritenour thinks it means "great big stream." Heckewelder also wrote that Unamis called it *Kithanne,* "big river." At least one settler along the Upper Delaware remembered in 1785 that some of his former Indian neighbors used what appears to be the pidgin or jargon word *lamasepose,* "fish creek" (*namées,* "fish," and *shiipoosh,* "creek"), when talking about the upper reaches of the river (Snell 1881:368). Early Dutch and English settlers adopted the latter name for a time in the form of the fairly common Dutch place name Fish Kill (e.g., Reading 1915:96–100) (not to be confused with present-day Fishkill in Dutchess County).

The name that the Munsees used for the river at the eastern end of their homeland before they were driven from the region has not been recorded. Heckewelder (1876:52) preserved a more westerly version, *Mohicanichtuk,* the Munsee word for "river of the Mahicans," used during later colonial times when most Munsees and many other East Coast refugees living west of the Appalachians were collectively called Mahikanders or Loups. Although Mahican probably came from their word referring to the tidal flow of the Hudson River, the name used by westerners evidently was based on *maengun,* an Ojibwa word for "wolf," noted as "loup" when used by French Canadians.

Etymology and American Indian Place Names

Etymologies provided by Whritenour for this volume go far in showing that a good number of Indian place names on present-day Greater New York area maps documented during colonial times come from the Munsee dialect of the Delaware language spoken by the region's first people. More than a few also come from the closely related Northern Unami dialect spoken by Indians who originally lived downriver in south-central New Jersey and neighboring Pennsylvania. Others come from the very closely related Wampano dialect spoken by eighteenth-century Moravian Indian converts from southwestern

Connecticut (identified as Quiripi in Rudes 1997). And a few names come from Shawnee, the Central Algonquian language of the Mississippi Valley refugees who lived at the Delaware Water Gap from 1694 to 1728.

The brief etymologies presented at the head of each entry in the first part of this book represent best guesses made by people who spoke, are conversant with, or study Munsee and other dialects of the Delaware language. Many readers will probably be surprised that etymologies put forward by such old standbys as Edward Manning Ruttenber (1906a), William Wallace Tooker (1911), and George P. Donehoo (1928) are mostly absent here. This is not meant as a gratuitous dismissal of their work. They and their contemporaries discovered and made available a great deal of information that remains useful to present-day students of local Indian history and culture. Those who follow them do well to stand gratefully on their shoulders. The fact that they are not intensively cited in this book does not mean that their works were not mined for the many good leads they still provide.

Present-day linguists, however, have found that nearly all homegrown etymologies, and inferences based on them, such as otherwise uncorroborated use of the translation of an Indian place name as "round" to prove that it was the name of a round-shaped lake, have not stood the test of time. This should not be surprising. None of these early writers worked directly with Delaware speakers familiar with the twists and turns of their language and its names. They instead depended on earlier word lists, some compiled by competent speakers of Delaware, such as the Moravians Heckewelder and Zeisberger, and others penned by writers far less familiar with Delawares or their language. Several of James Hammond Trumbull's (1881) attested etymologies from New England Eastern Algonquian languages, still spoken in parts of Connecticut when he was working, continue to be highly regarded and are cited here. Trumbull and his contemporaries did the best they could with what was available to them. Any shortcomings arising from the fact that they worked at a time before modern linguistics completely transformed our understandings of human speechways cannot be laid at their feet.

Many Indian names from New England and other places crossed into the Greater New York area. The number of these imports far exceeds those invented or mistakenly identified by local people as Munsee or Delaware place names. Most, such as Sioux, Iroquois, and other tribal names, as well as state names, even Wyoming, which comes from the Munsee language, have no direct or deliberate connection with Greater New York's first people while they lived in their homeland. Although the combined total of Indian names from all sources on regional maps is considerable, they still make up only a small portion of the region's current stock of toponyms.

Not all of the words in this book come directly from Indian languages. Some are much-altered anglicizations, such as Lamington, which can only be connected to its original Indian name Allamotunk through research. More than a few of these words are English or Dutch names for Indian places. Some, such as Fort Neck, are fairly straightforward. Others demand travel along more twisty paths to arrive at their Indian linguistic origins. Prescott Brook, for example, is twice removed from its Indian beginnings. Whritenour shows that the name is a fairly recent anglicization of *Piskot,* a colonial Jersey Dutch word meaning "skunk or polecat." Colonists, it turns out, evidently bestowed the nickname on a local Indian whom John Reading, Jr. (1915:45), unsuccessfully tried to engage as a guide in 1715 during one of his surveys of West Jersey.

The Influence of European Colonists on Place Names

The Europeans who colonized Munsee and other Native northeastern coastlands between Delaware Bay and Boston Harbor chose not to chronicle Indian names for many of the region's strategically significant waterways. In fact, the Connecticut and Housatonic rivers and Narragansett Bay are about the only major bodies of water in the region that retain their Indian names (Chesapeake, Susquehannock, and Potomac lay beyond the area; even there, Virginians called what was the most important river to them the James in honor of their king). Colonists instead used words from their own languages, such as Hudson, to support their nation's right to control lands watered by the region's other major rivers and bays.

Europeans used their own words for the Hudson River from the beginning. Henry Hudson himself referred to it as the River of the Mountains when he first sailed up into it in the fall of 1609. Those who followed him called it the Mauritius River (after Prince Maurice of the Dutch royal house of Orange), the North River (its location at the northern end of the Dutch colony of New Netherland), the Manhattan River (its only attribution of Munsee origin), and, finally, the Hudson River. The Iroquois called it *Cahohatatea* (river of the mountains below the Cohoes Falls at the head of navigation of the Mohawk River above Albany); *Ohio,* their word for "beautiful river"; and *Skanetade,* a Mohawk word meaning "the other side" (all in Grumet 1981:22).

Settlers also tended to apply their own names to places they prized the most. Those Indian names not stored away for later use in title defenses during litigations that historian Francis Jennings (1975:128–45) called "deed games" were largely relegated to more remote and much less desirable stone-choked streams, mosquito-infested swamps, sandy barren lands, and forbidding rocky mountain summits. Such placements almost certainly metaphorically represented colonial expectations that actual Indians surviving disease, despair, and dispossession would occupy correspondingly secondary positions in the maps, societies, and memories of settlers who hoped to supplant them. Other Indian toponyms for major rivers and places farther in the interior that remained on the map, names such as Mississippi, Ohio, Illinois, and Michigan, came relatively late into English from French sources. They and others like them seem to have found secure places on North American maps only after they became trophies marking the British triumph over the French and their Indian allies after the end of the Seven Years' War.

What Are Indian Place Names Doing on Modern Regional Maps?

As mentioned above, most Indian names documented by colonists were preserved in archives to secure titles and provide evidence in land disputes. The small percentage that made it onto colonial maps probably initially got there in large part because settlers had to share the same space with Indians for a very long time. No matter what they thought or felt about Native people, colonists had to adopt place names that were mutually acceptable to them in order to find their way through the region's cultural landscape as long as Indians remained a force to be reckoned with in their homeland.

But what about the years after the colonists finally drove the Indians from their ancestral lands? As mentioned earlier, endless repetition in official records may have all but stripped away the Indian affinities of names such as Manhattan. Similarities, such as the resemblance of the German word *minnesinger* (a kind of medieval balladeer still familiar to the many descendants of German settlers in the Upper Delaware Valley) to the Indian name Minisink (often spelled the same way as the otherwise unrelated German word), doubtless eased transfer of other Indian names onto colonial maps. Sheer disinterest, on the other hand, apparently preserved those fixed onto tiny or remote streams, creeks, and rivulets. Numbers of other Indian names dwindled to near extinction, especially in places where relations between natives and newcomers during the last years of contact had not been particularly cordial.

Following independence, many Americans doubtless expected that Indian place names, like the people themselves, would gradually vanish from the landscape. Even missionaries devoted to winning Native converts did their work with the undisguised goal of ending their cultural lives as Indians. Everyone thought that Indian names would be replaced by a progression of new American names symbolically reflecting the nation's cultural and physical transformation. Even those retained as trophies might ultimately be supplanted by names celebrating the new nation's triumphs on battlefields and at treaty tables.

At first, it looked as if this was the way things would happen. Indian names disappeared from most of the places where they were recorded. Names such as New Netherland and New York proclaimed their new places as centers of old-country rebirth and regeneration. Families who did the hard work of seeing new generations into existence gave their surnames and places of origin to streets and neighborhoods in cities. Others did the same in rural towns, villages, mill seats, and crossroads. The practical utility of familiar names telling people where family businesses ground grain or sold goods (the word "store" still appears as a part of place names in sections of rural Virginia) is self-evident.

Indian place names, however, like the people who first gave them breath, refused to vanish. Instead of decreasing, the number of Indian place names on American maps everywhere began to increase even before Indian populations began to rebound from their low point around 1870. New names were required as the old family-centered social order gave way to an altered cultural landscape dominated by corporations, cities, and suburbs. The U.S. Post Office's efforts to help meet these needs stimulated increases in numbers of familiar, resurrected, imported, and invented Indian place names given to their branches. The total number of these offices alone more than tripled from 18,417 in 1850 to 62,401 in 1890 (Historian, United States Postal Service, 2008). Determined to bring some order to the often chaotic debates surrounding selection of official names for these and other federal properties, President Benjamin Harrison issued an executive order for the creation of a U.S. Board on Geographic Names on September 4, 1890. States and many local municipalities soon began following suit, setting up committees of their own to regulate selection, form, and placement of names on their maps.

Those interested in post offices were not the only Americans who began restoring Indian names as they looked back nostalgically to past glories as the halcyon days of America's formative decades faded into memory. Novels such as James Fenimore Cooper's *Last of the Mohicans* (1826) and, even more notably, poems such as Henry Wadsworth Longfellow's *Song of Hiawatha* (1855) dutifully memorized by generations of American schoolchildren helped pave the way. They and others like them satisfied romantic longings for a nobler and more exciting past by supplying new images of tragically inspirational Indian heroes and heroines. Gallant resistance against overwhelming odds led by patriot chiefs such as the Ohio Valley Shawnee leader Tecumseh and the Seminole war captain Osceola assured that their names and others like them were immortalized on American maps.

Numbers of Indian place names increased on American mapscapes along with more blandly nondescript names of places such as Plainview and Pleasantville as a growing sense of aesthetic appreciation for the more sublime aspects of nature swept across the nation. Finding few verifiably authentic Indian names on regional maps, local historians such as Ruttenber and Tooker stormed the archives to satisfy the demand. Old names forgotten for decades suddenly reappeared everywhere. New findings were not enough, however. Chamber of commerce boosters vied with local antiquarians to give newly imported, improved, and invented Indian names to freshly constructed camps, schools, and the multitude of artificial lakes, ponds, and reservoirs whose waters rose behind dams piled up across rivers and streams at nearly every likely locale in the region. Spellings identical or nearly so to forms found only in books reveal the origins of many of these names.

It helped that Indian names, especially those from Algonquian languages such as Munsee, tended to possess a certain mix of hard and soft sounds that seemed to fall with particular melodic sweetness on American ears. Their origins as words from ancient aboriginal languages added just enough of a sense of the unfamiliar to give them an air of romantic mystery, albeit an unthreatening one, since the Native nations whose people first breathed life into the words (and in so doing staked their claim to the places they named) were thought to be far away or extinct and no longer interested in taking back their ancient lands. Americans putting Indian names back on maps nevertheless often took care to change them ever so slightly from their original spellings in colonial documents. Most were tweaks such as adding or subtracting a vowel or two somewhere in the name or transforming a soft-sounding English consonant such as *c* into what was regarded as a harder, more Indian-sounding *k*. However the changes were made, all seem to have been undertaken to raise the pedigrees of the names and deflect any possible negative side effects by disguising, even if only just barely, their derivative origins from different cultures in old documents.

Employment of such subtle subterfuges shows how gingerly people handle words such as place names often regarded as repositories of metaphorical power and symbolic authority. Names in general and place names in particular that verbs and their allies try to fix in time and space seem to defy restraint. Most are multitaskers that can mean different things to different people at different times in different places. Those not glorying in the name of a battlefield, for example, can choose instead to remember the old church, mill, village, or Indian town that gave its name to the site. To complicate things further, places can retain old names that have little or nothing to do with changed roles and functions.

This kind of multitasking, often referred to as multivocality, can produce a sense of uncertainty that scholars like to call liminality. People, especially those interested in asserting their authority, can use such words to appropriate meanings that symbolically express the extent of their control and metaphorically deny power to others. Viewed this way, the words on present-day maps are not, in the strictest sense of the term, Indian names. They are instead American versions of Indian names selected in accordance with American values and maintained to serve American interests.

Problems of Translation and Identification

A colleague of mine once penciled in the Italian expression *traduttore, traditori* ("translator, traitor") beside a passage in a prepublication draft of a paper he was kind enough to look at for me. It was his way of pointing out what can happen when trying to make sense of the meaning of words in foreign languages. These problems are magnified considerably when those words rise up from the grave, as it were, on old manuscript pages. Accidents of time, cross-cultural miscalculations, semantic differences (distinguishing damnation from nation of dames or dams, for instance), and the near impossibility of recognizing different meanings attached to identical-sounding words—toad and towed, or wheel, weal, and we'll—when they are divorced from their original contexts are only some of the traps lying in wait for the most careful philologists. Uncritical website writers and local enthusiasts often do little more than perpetuate old mistranslations while adding a few new ones of their own. No wonder so acute a student of place names as Ren Vasiliev (2004:235) had to figuratively throw up her hands in frustration in her entry on the Munsee place name Wawarsing, "said to mean," she wrote, "'black birds' nest' or 'holy place of sacred feasts and dances' or neither."

The people who first uttered the place names in this book did not use a written language of their own to record what they meant by them. Although many Indians used the region's pidgin trade jargon, and a few learned to speak one or another of the languages spoken by settlers, literacy was rare among Indians. Indian students in the United States and Canada initially schooled in mission classes began writing full-dress histories of their own only during the nineteenth century. Because of this, all Indian words originally recorded in the colonial Greater New York area come from records penned by people who did not really understand Indians, their languages, or their histories. Reliance on the earliest orthographies or transcriptions that seem to conform to phonological or syntactic rules cannot by themselves guarantee that a name was written down exactly as it was spoken. Comparison of the wide range of spellings used to record particular names suggests that the dropping, adding, and reshuffling of sounds was a frequent occurrence. Appearances of compound names combining parts drawn from different languages, such as Hobocanhackingh (the name of the Belgian city of Hoboken resembling a Munsee word for pipe attached to *hackingh*, a pidgin Delaware word for "place") and Manhatesen (a Dutch plural tacked onto the end of a Munsee word), further muddy etymological waters. Surviving documents recording sounds that scribes tried to reproduce, moreover, tend to be derivative transcribed copies of long-lost originals. The majority of these records are fragmentary, incomplete, far removed from their original cultural settings and historical contexts, and often riddled with internal inconsistencies and contradictions.

Raymond Whritenour, a particularly scrupulous student of Munsee and Northern Unami names and naming and editor of a concordance of Zeisberger's Delaware-English dictionary (Whritenour 1995), who has supplied most of the Munsee and Northern Unami etymologies presented in this volume, uses a rigorous winnowing process to narrow the field down from the preposterous to the possibly authentic. He employs five quintiles represented by percentages to express the level of confidence he assigns to each translation. These range from 99 percent (near certainty), 75 percent (highly probable), 50 percent (maybe, maybe not), and 25 percent (possible, but unlikely), to 1 percent (next to impossible). In his analysis of Kevin Wright's (1994) translations of seventy-four northeastern New Jersey Indian place names (nearly all of which he finds inadequate), for example, Whritenour uses his system to express his degree of confidence in his etymologies. Only nine of these translations make it into his 99 percent quintile, ten are included in the second, and eighteen inhabit the third rung. By his own reckoning, fewer than half of the retranslations Whritenour made for this book have a 50 percent or greater chance of being correct, and only 12 percent could confidently be thought to approximate their original meanings to a "near certainty." I have adopted many, but not all, of his suggested etymologies, culling out those he thinks have less than a 50 percent chance of approximating original meanings. Unless otherwise cited, all mentions of his work acknowledge these contributions to this volume. Translations drawn from Brinton and Anthony (1888), Heckewelder (1834), Trumbull (1881), and Zeisberger (1887) are identified but otherwise uncited to avoid tedious repetitions of references already well known to readers interested in them. Translations of Southern Unami dialect place names identified by the late Eastern Oklahoma traditionalists Lucy Parks Blalock and Nora Thompson Dean as Lenape words come from contributions originally published in Boyd (2005) and Kraft and Kraft (1985:45), respectively.

Also unmentioned are the innumerable questionable, erroneous, and downright spurious translations readily encountered in any online query whose sheer volume and vast numbers would make this book unpublishable. This also spares readers endless repetitions of negative statements refuting etymologies tracing place name origins to undocumented Indian sachems, imaginary features, and legendary occurrences. To keep things as clear as possible, I have done everything I can to document the original primary sources for those names presented in this book. Particular care has also been taken to assure that the Indian words themselves are presented as they appear in the original records. Proper nouns are only standardized when inconsistently or ambiguously employed in secondary sources such as genealogical reconstructions and historical treatises. I identify as Delaware, for example, all Lenape words Heckewelder did not distinguish by dialect in his 1834 place name monograph.

I have also modernized archaic or idiosyncratic spellings of words other than names to minimize distractions caused by otherwise insignificant variations produced before grammarians standardized the English language. All orthographies of Munsee words follow the simplified format used by linguist John O'Meara (1995) in his Munsee-English dictionary. Asterisks (*) precede philological reconstructions not appearing in historical documents. Unless otherwise cited, information on Munsee people and polities in both parts of this book comes from my survey of Munsee history (Grumet 2009). The often-mentioned 1770 William Scull, 1792 Reading Howell, and other early maps available on the Library of Congress website are also uncited. I have noted information essential to understanding the history and meaning of place names stored in archives, published in books and articles, and readily available on the Internet. Listings in the National Register of Historic Places are included partly because such designations provide handy guides to historic places significant in American culture and history and also because the National Register's computerized Information System is a readily available source for carefully vetted documentation. Less essential secondary data such as dates of town or borough incorporation or construction of rail lines found on the Internet are not cited here.

Structure and Organization

This book is divided into two sections. The first contains entries summarizing information relating to the nearly 340 Indian place names preserved in colonial records. All of these names first appeared in records written by Europeans between 1609, when a crewman on Henry Hudson's ship first recorded the name Manahata, and 1801, when most of the first people remaining in their homeland were forced to leave their last communities in the region and join the westward trek that ended for most in exile communities in Ontario, Wisconsin, Kansas, and Oklahoma.

Names listed in the first part of this book represent those that can reasonably be fixed in time and space on securely dated documents. Confidence in their authenticity as Indian names increases when they appear embedded in formulaic statements entered onto deed texts, survey returns, or court depositions affirming that a certain place name is called thus and such by the Indians. Philological verification affirming Indian etymologies for such words represents the gold standard for authenticating Native place names in this region.

The tiny number of Indian place names meeting this etymological gold standard is hardly enough to fill the pages of a brief article, let alone those of a book-length study. The entries herein primarily focus, therefore, on the many ways Indian place names documented during colonial times fit into the history of Greater New York. Although we cannot be sure of their meanings or original functions, we can determine how and when colonists and those who followed them recorded particular names of places. Examining this information over time and space can reveal some of the ways these people used, stored, forgot, rediscovered, and, like cartographic resurrectionists, restored long-forgotten Indian place names from the region back onto present-day maps.

The book's second section presents nearly two hundred invented or imported place names regarded as local Indian names at one time or another. This group of misnomers includes two perpetrated by me, Hassokie and Hawtree (archaic English words for "tufted grass" and "hawthorn tree," respectively), and another pair, Jamaica and Mosholu, that I had a hand in perpetuating. I take no little pleasure—well, not so much that you would easily notice—in debunking the remaining sacred cows in the second part of this book. Although the results will not please many who have faithfully believed in traditions associated with some of these names, others who prefer their sacred cows served straight up will, I hope, find the information useful. The effort taken to cull such cattle from the herd should pay off in several ways, not least by helping readers distinguish folk facts and historians' suppositions from verifiably dated and located documentary data from colonial times.

Place names in both sections of this book are alphabetically arranged according to the ways they are most commonly spelled today. Each is highlighted in bold-faced type, and all are located by county and state. Every entry includes information identifying the name's language of origin when known and a brief discussion of its more credible etymologies if such exist. Silence rather than tedious repetition of negative findings is also observed here to indicate when a name has no demonstrably verifiable etymology or historic association. The current status and location of each name is then examined. The earliest known occurrence in written records is noted and the history of each name's appearance on regional maps is recounted. Source materials used in this survey are noted with parenthetical citations. Colonial spellings of common nouns found in historical records have been modernized, and colonial-era dates have been adjusted to New Style whenever possible. The main data sections of this book are preceded by the introduction you are now reading, a timeline presenting key dates and events, and maps showing present-day locations of place names mentioned in the text. The data sections are followed by source lists that separate manuscript repositories and published works.

I hope that more scholarly inclined readers will use the material in this book to test hypotheses that can contribute new insights into what North American Indian place names teach us about American history in particular and what they do for people in general. In this way, they can do more than help us gain clearer understandings of where, when, and why Indian names appear and disappear. Clearer understandings of the biographies of place names can help future writers get at bigger questions, such as why people often feel so passionately about place names, and why names of places important to otherwise much-reviled, long-exiled, and mostly forgotten people remain on present-day maps in their former homelands at all.

Part 1

COLONIAL-ERA INDIAN PLACE NAMES

ACQUACKANONK (Passaic and Sussex counties, New Jersey). Heckewelder thought Acquackanonk sounded much like a Delaware word, *tachquahacannéna*, referring to a place where people made pounding blocks from *tachquahcaniminschi*, "gum trees." Several writers have found another Delaware word, *achquanican*, "fish dam," recorded in Brinton and Anthony (1888:11) a plausible possibility. Whritenour thinks Acquackanonk sounds almost exactly like a Munsee word, *axkwaakahnung*, "at the place of lampreys." Acquackanonk was originally a general name for the tidewater section of the Passaic River at and around the present-day cities of Passaic and Clifton. The word first appeared in colonial records as Gweghkongh and Hweghkongh, the home of several sachems signing the July 15, 1657, deed to Staten Island (Gehring 2003:141–42). Three deeds to places in and around the city of Passaic signed between April 4, 1678, and April 9, 1679, identified land there as Aquickanucke, Haquequenunck, Aquenongue, and Aqueguonke (Budke 1975a:47A–47E). Colonists quickly adopted Acquackanonk as the name for the area. They retained it when they incorporated their community as a township in 1693 only to abandon it in favor of another Indian name, Passaic, when they voted to establish a city government at the locale in 1873. Popular during the nineteenth century, the name gradually fell out of favor during the early 1900s. Today, it appears on regional maps as the name of a neighborhood in the city of Passaic and as the name of a YMCA camp and its lake in the borough of Ogdensburg in Sussex County.

ALEXAUKEN (Hunterdon County, New Jersey). Nora Thompson Dean identified Alexauken as a Southern Unami word, *alàxhàking*, "barren land." Whritenour thinks it sounds very much like a Munsee word, **eeliikwsahkiing*, "in the land of ants." Alexauken Creek is a small Piedmont stream that today forms a boundary between the Hunterdon townships of West Amwell and Delaware. The creek rises near the Delaware Township hamlet of Headquarters and flows into the Delaware River just north of the city of Lambertville. Alexauken was first identified as Hockin Creek in an April 16, 1701, entry for a survey of three thousand acres of West Jersey Society land "near Wishilimensey" (Whitehead et al. 1880–1931[21]:388). An "Indian path leading from Itshilominwing unto Noshaning" near "a small brook or run having its rise or first spring about Achilomonsing" was next mentioned in an Indian deed in the area signed on June 5, 1703 (New Jersey Archives, Liber AAA:443–45). The name subsequently turned up in 1742 in another surveyor's notebook as the Alequemsokm Mill, near the mouth of the creek (Moreau 1957:13), and as Aliashocking Creek on the 1770 William Scull map. Later spelled Alexsocum and Alexsocken (Gordon 1834:93), the name is now used to identify the small hamlet in Delaware Township near the stream's mouth where colonists first settled in 1695, the road paralleling that stream, and the 690-acre State Wildlife Management Area acquired in 2001 to regulate development on protected land on the West Amwell side of the creek.

ALLAMUCHY (Morris, Sussex, and Warren counties, New Jersey). Nora Thompson Dean thought that the present-day spelling, Allamuchy, most resembled a Southern Unami word, *alemuching*, "place of cocoons." Whritenour thinks its earliest spelling, Allamuch-Ahokking, more closely resembles "at the land of cocoons." Allamuchy presently is the name of several municipalities, a pond, a mountain (elevation 1,222 feet), a state park, a natural area, a Boy Scout reservation, and a number of other places located in a part of the New Jersey highlands that divides the Musconetcong and Pequest river drainages. Most of these places lie on one side or the other of the border between Byram Township in Warren County and Allamuchy Township in Sussex. Others, such as Allamuchy State Park, extend across both borders into Morris County. Surveyor John Reading, Jr. (1915:42), wrote down the earliest known references to the name in notes entered into his journal during the month of May 1715. Recording the words Allamuch-Ahokking and Allamucha, he noted that both were names of an "Indian plantation upon a branch of Pequaessing [Pequest] river" just west of a vantage point where the Delaware Water Gap could be seen in the distance. Although a few Quakers had settled in the area by 1764, most colonists were not drawn to the area's rock-strewn uplands and malaria-ridden Great Meadows (a former glacial lake, one of whose earlier names, "Shades of Death," now adorns a local road). Intensive settlement in the area began only after railroads brought flatcars carrying steam-driven pumps capable of draining the local swamp lands. The same railroads carried produce grown in the fertile expanses of black dirt drained by the pumps to markets in New York and metropolitan New Jersey. Local farmers resurrected Reading's Allamucha in the slightly altered form Allamuche as the name for their agricultural community and the nearby lake and mountain by the time Gordon (1834:92–93) published his gazetteer. Residents gave the name to the post office opened at the village sometime before 1855 and later bestowed it in its present form on their township following its organization in 1873.

Developer John Rutherford Stuyvesant acquired large blocks of land in the area at the turn of the twentieth century. He built one of his mansions on the banks of Allamuchy Pond and named another part of his estate Allamuchy Farms. The state of New Jersey used Green Acres bonds to acquire much of this land during the 1960s. The 8,683-acre Allamuchy Mountain State Park established on this tract is currently managed by the New Jersey Division of Parks and Forestry. A 2,440-acre area originally set aside by Stuyvesant as a game preserve was among the first tracts designated as a State Natural Area in 2010. Another place bearing the name, the Allamuchy Freight House, which was built in 1906 by the Lehigh and Hudson River Railway at its station in Allamuchy Township, was listed in the National Register of Historic Places in 2002.

AMAWALK (Westchester County, New York). James Trumbull (1881) thought that Amawalk derived from a Delaware word, *nam'e-auke*, "fishing place." Amawalk is presently the name of a village, a dam, a hill, and a Friends Meeting House (built in 1831 and listed in the National Register of Historic Places in 1989) in the town of Somers. The original village of Amawalk now lies beneath the waters impounded by the Amawalk Dam built across the Muscoot River in 1897 as part of New York City's Croton Reservoir system, about three miles northwest of the place where the river falls into the Croton Reservoir. Amawalk was probably the location of the Ammawaugs Indian town mentioned as being "on the east side of Hudson's river a little below the highlands" in 1720 (Connecticut State Archives, Indian Papers Series, 1:92a–92b). Ammawogg was subsequently mentioned as the home of the Indians who sold most of the last lands remaining in Indian hands in present-day Ridgefield, Connecticut, on July 4, 1727 (Hurd 1881:636–37). Ruttenber (1906a:34) traced Amawalk's origin to the even earlier place name Appamaghpogh, noted in the August 24, 1683, deed to land in present-day Somers, mentioned in Robert Bolton's (1881[1]:86–87) history. A man identified by the similarly spelled name Appamankaogh signed the December 26, 1652, deed to land at the mouth of the Raritan River (New Jersey Archives, Liber 1:9). He may have been the same person identified as Oramapouah in the November 22, 1683, deed to land at White Plains abstracted by Robert Bolton (1881[2]:536).

AMBOY. See **PERTH AMBOY**

AMOGERONE (Fairfield County, Connecticut). Amogerone was noted as one of two sachems of Asamuck signing the July 18, 1640, Indian deed for land in the present-day town of Greenwich (Hurd 1881:365–66). The name was revived in 1876 by the still-functioning Amogerone Volunteer Fire Company Number 1 and has more recently been given to the Amogerone Cross Way (formerly Amogerone Place), a nearby one-block-long byway in the Greenwich town center.

ANALOMINK (Monroe County, Pennsylvania). Heckewelder wrote that Analomink came from a Delaware word, *Nalamáttink*, "the place where silk worms spring up, or mount; silk worms' place," in reference to the mulberry trees that grew abundantly at the locale. Analomink Lake and village are in Stroud Township. The name first appeared in colonial records as Manawalamink Island (present-day Schellenbergers Island at the mouth of Brodhead Creek) in 1733. Lehiethaes Creek noted just north of the Delaware Water Gap in the 1738 Evans map of the Walking Purchase (in Gipson 1939) was soon renamed Analoming Creek. Daniel Brodhead, an Indian trader and militia officer originally from Marbletown, New York, purchased land at the lower part of the creek in 1737. He gave his first name to Danbury, the place where he built his home, and gave his family name to Brodhead Creek, the stream that ran by it. Then as now, the creek rises above the present-day hamlet of Analomink and flows southward past East Stroudsburg to its junction with the Delaware River at the village of Delaware Water Gap. A house Brodhead built for the convenience of passing Moravians in 1744 became the nucleus of the Danbury Indian mission community. The latter locale was abandoned after an Indian raiding party burned it in 1755. Residents of the hamlet of Spragueburg, which sprang up around the millpond that rose behind the dam built across an upper part of Brodhead Creek in the 1830s, later called it Analomink Lake. Finding the lake's Indian name more appealingly unique than the relatively colorless Spragueburg, they finally formally adopted Analomink as the name for their community in 1905.

APSHAWA (Passaic County, New Jersey). Whritenour suggests that Apshawa sounds like a Munsee word, *ahpchuwal,* "upon the mountains." Apshawa Lake, Brook, and locality are located in West Milford Township. The name appeared in a survey return dated July 14, 1790, for land "situate[d] about one-half mile from Pequonnock River on a brook called Andersons Brook on the southeast side of Apshawaw Mountain" (New Jersey Archives, East Jersey Board of Proprietors Papers, Patent Book[B]:107, Survey Book[S-8]:454). Today's Apshawa Brook may be the same Andersons Brook, which currently flows into the Pequannock River at the hamlet of Smith Mills just west of the borough of Butler. Paterson businessman John W. Hennion brought Apshawa to wider attention when he expanded an earlier impoundment of brook waters that had served as the focus for a local private camp. His Lake Apshawa Fishing and Hunting Club became a popular tourist destination whose clientele included President William McKinley (Nelson and Shriner 1920[1]:124–25). That Lake Apshawa is now called Henion Pond. The Butler Water Company acquired the present Lake Apshawa in 1912. Christening it the Apshawa Reservoir, the company provided lake water to municipalities in the area until the 1970s, when it sold the lake (retaining its water rights) to Passaic County. County officials ultimately joined forces with the New Jersey Conservation Foundation to jointly administer the Apshawa Preserve, recently established on 576 acres of land surrounding the lake.

AQUASHICOLA (Carbon and Monroe counties, Pennsylvania). Heckewelder thought Aquashicola sounded much like the Delaware word *achquaonschícola,* "the brush-net fishing creek; the creek where we catch fish by means of a net made with brush." Aquashicola Creek parallels the northern scarp of Kittatiny Mountain from its headwaters near Wind Gap in Monroe County west past its junction with Buckwha Creek at Little Gap in Carbon County. From there it flows through the hamlet of Aquashicola to the city of Palmerton, where it joins the Lehigh River just above the Lehigh Gap. Moravians began visiting the Indian town of Meniolagomeka, located along the creek's upper reaches, in 1743. They gradually established a mission at the site. The place name Aquanghekalo first appeared at the locale in a deed dated 1749 (Hazard 1852–60[2]:33). Both the missionaries and their Native converts were compelled to abandon the town when Delaware and Shawnee warriors, allied with the French in their war with Great Britain, began ravaging settlements in the area in 1754. The stream subsequently was noted as Aquanshicola Creek on William Scull's map of 1770. Americans moving to new homes built near the creek's lower end in 1806 named their community Millport.

Later in the century residents changed the community's name and that of the local post office opened in 1855 to Aquashicola after resorts opened along the creek's upper branches helped establish the place's reputation as a Poconos resort destination.

AQUEHONGA (Sullivan County, New York). Whritenour finds that Aquehonga sounds much like the Munsee word *aakawahung*, "that which is protected from the wind." The name of "the River Aquehung or Bronxkx" mentioned in the March 12, 1663, deed to land in the West Farms section of the Bronx abstracted in Robert Bolton's (1881[2]:433–34) history may share a similar etymology. Aquehonga is currently the name of a camp in the Ten Mile River Boy Scout Reservation that until recently catered almost exclusively to troops from Staten Island. The name first appeared as Eghquaons, the Indian name for Staten Island, in the deed documenting the second sale of the place signed on July 10, 1657 (Gehring 2003:141–42). The island had first been sold by Indians in 1630 before being repurchased by settlers who had twice been driven away by Indian warriors, first during Governor Kieft's War in 1641, and again during the Peach War in 1655. The island was subsequently identified as Aquehonga Manacknong in the deed sealing the third and final Indian sale of Staten Island concluded on April 13, 1670 (Palstits 1910[1]:338–43). Lingering hard feelings over the circumstances surrounding these sales may explain why Staten Island is the only county in the Munsee homeland that does not have a local Indian place name on its map. Borough residents have not, however, totally neglected the only Indian place name associated with their island documented in colonial records. Aquehonga became a popular name adopted by Staten Island social clubs and fire brigades during the late 1800s. It survives today as an internal migrant in the Munsee homeland some one hundred miles distant from its place of origin.

ARMONK (Westchester County, New York). Whritenour thinks that Armonk most closely resembles a Munsee word, **alumung*, "place of dogs." Armonk is the name of a hamlet located in the town of North Castle. Dutch colonial official Cornelius van Tienhoven penned the earliest known reference to a rivulet he identified as Armonck situated somewhere between the East and North (today's Hudson) rivers on March 4, 1650 (O'Callaghan and Fernow 1853–87[1]:366). Settlers moving into the area during the 1700s regarded it as the original name for the Byram River, a stream that rises some miles above the present-day hamlet before it veers eastward into Connecticut, where it debouches into Long Island Sound along the state line at Portchester, New York. The village of Armonk was known as Sands Mill when the local postmaster, at the suggestion of Westchester historian Robert Bolton (1881[1]:2), adopted the present name for the post office opened at the locale in 1851.

ASHAROKEN (Suffolk County, New York). Today, Asharoken is the name of an incorporated village and several localities on Eatons Neck in the town of Huntington. Asharoken was originally the name of an influential Matinecock sachem who placed his mark onto a substantial number of deeds to lands along Long Island's north shore in the present-day towns of North Hempstead, Oyster Bay, and Huntington between 1646 and 1669. The area was already known as Eatons Neck (after Theophilus Eaton, governor of the New Haven colony that was home to many of the village's first English settlers) when Asharoken signed the July 30 and 31, 1656, conveyances to land in the area (C. Street 1887–89[1]:6–7; Palstits 1910[2]:403–405). A businessman named William Codling selected the sachem's name exactly as it was spelled in the 1656 deeds to adorn his Asharoken Beach development built around the turn of the twentieth century. Residents intent on controlling local services formally incorporated the community as the village of Asharoken in 1925. Use of the name has since expanded to include a number of natural and cultural features in and around the village.

ASHOKAN (Ulster County, New York). Whritenour suggests that Ashokan sounds like the Munsee word *aashookaan*, "people are walking in the water." Today, Ashokan is the name of a dam, a reservoir, and a hamlet in the Catskills. A passing reference to a path leading from a Rochester town farmer's residence "to Ashokan" set down on October 8, 1706, represents the earliest known notice of the name (Brink 1910[10]4:100). Other records show that colonists began building settlements in and around the area by the 1730s. The local post office bore the name Ashocan for two months before the postmaster changed it to Caseville in 1832. Ten years later it was changed again, this time to Shokan, its present name (Kaiser 1865). In 1907 New York City officials planning construction of a major reservoir in the area, soon given the name Ashokan, began acquiring land along the Esopus Creek in the towns of Hurley and Olive. The 123-billion-gallon impoundment (the second largest in the New York City Water Supply System) completed in 1917 required relocation of the village of Ashokan and several other settlements to higher ground. Today, the hamlet of Ashokan lies just north of the reservoir along New York State Route 28 in the town of Olive. Musicians Jay and Molly Ungar's composition "Ashokan Farewell" featured in the 1990 Civil War miniseries on PBS has broadened awareness of the name beyond the region's borders.

ASPEN (Fairfield County, Connecticut). Based on Asproom, a presumably Wampano or Quiripi word, Aspen Mill Road is located in the town of Ridgefield. During the 1960s the local planning commission saw to it that this place name would win any prize awarded to the most thoroughly disguised Indian place name in the region when it insisted that a local developer adopt Aspen instead of Asproom, the colonial-era name he had chosen. Asproom made its first appearance in town records as a mountain mentioned in the November 22, 1721, Indian deed to land in the area (Teller 1878:23–24). Colonists subsequently gave the name to a local meadow, a swamp, and a bridge (Sanders 2009). These places were later given other names, and Asproom disappeared from local maps. Today, Aspen is the only name on the map providing any indication that Asproom was once a widely used Indian name in Ridgefield.

ASPETUCK (Fairfield and Litchfield counties, Connecticut; Orange and Westchester counties, New York). Whritenour proposes that Aspetuck sounds a great deal like a Munsee word, *uspahtung*, "an incline or hillside place," a translation very similar to a Southern Unami etymology suggested by Nora Thompson Dean. Aspetuck is a place name found in several locales in and around the region. It first appeared as the Aspatuck River mentioned in an April 11, 1661, boundary dispute agreement involving the Sasqua and Norwalk Indians in the present-day town of Fairfield (Wojciechowski 1985:96). The river, also called Great Brook during later colonial times, is a seventeen-mile-long stream that rises at Flat Swamp in the village of Newton. From there it flows south into the Aspetuck Reservoir and past the village of Aspetuck to its confluence with the Saugatuck River just north of the town of Westport. Aspetuck Neck at the mouth of the Saugatuck River was sold separately in a deed dated January 19, 1671 (ibid.:98). That same year, John Wampus, a Nipmuck Indian from eastern Massachusetts, claimed land at Aspetuck by right of his daughter's marriage to Nowenock, a local leader from the Hudson Valley. Farther north, the separate ten-mile-long East Aspetuck Creek rises at Lake Waramaug and flows south below the Aspetuck Ridge into the Housatonic River at the Litchfield County village of New Milford.

Today, many scholars refer to the Native people who lived along the Aspetuck River in Fairfield as Aspetuck Indians. People in the county began giving the name to roads, ridges, and a wide variety of organizations, businesses, and housing developments located in a triangular area bounded by Redding, Bridgeport, and Darien during the mid-1800s. Aspetuck village is now known as the place where Helen Keller made her home. Her house is one of many carefully preserved early buildings included in the Aspetuck Historic District, listed in the National Register of Historic Places in 1991. The similar-sounding name Aspetong appeared in an August 24, 1674, deed in Monmouth County, New Jersey (New Jersey Archives, Liber 1:271[68]–270[69]). An identically spelled version of the name retrieved from old Connecticut records became very popular in and around northern Westchester County during the late 1800s. It survives there today as Aspetong Road in Bedford, Old Aspetong Road in Katonah, Mount Aspetong in the town of North Salem, and the still-remembered though since subdivided Aspetong Farm estate built in 1899 just across the Hudson River in the Orange County community of New Windsor.

ASSISICONG (Hunterdon County, New Jersey). Whritenour thinks Assisicong sounds much like the Munsee word *asiiskoong*, "place of mud." Nora Thompson Dean thought that Assiscunk, the name of a different creek farther south in Burlington County, sounded very much like a Southern Unami word, *asiskung*, "muddy place." Assisicong Creek is the name of a creek in Raritan Township that rises in the hamlet of Cherryville and flows southeast for three miles to its junction with the South Branch of the Raritan River. The twenty-six-acre Assisicong Marsh Natural Area lies across from the mouth of the creek in Readington Township. The Natural Area is managed as a component of the more than one-thousand-acre South Branch Reservation system administered by the Hunterdon County Department of Parks and Recreation. The name first appeared in 1703, first in the forms of Asiukowoshong and Asunowoshong in the June 5 deed to land between the Neshanic River and the South Branch of the Raritan (New Jersey Archives, Liber AAA:443–45), and then as Asurhwoorkong in the November 11 deed to land between the South Branch and Delaware rivers (ibid.:434–35). Six years later it was mentioned on October 7, 1709, as the "run or brook called Asiskowectkong" forming the northern boundary of a property immediately south of the aforementioned tract (ibid., Liber BB:323–24).

ASSUNPINK (Mercer and Monmouth counties, New Jersey). Nora Thompson Dean (in Boyd 2005:444) identified Assunpink as a Southern Unami word sounding much like *ahsen'ping*, "rocky place that is watery or a place where there are stones in the water." Assunpink is presently the name of a creek, a dam, a lake, a wildlife management area, and much else in central New Jersey. The twenty-three-mile main branch of Assunpink Creek rises in Monmouth County, where it enters the New Jersey Department of Environmental Protection's 6,324-acre Assunpink Creek Wildlife Management Area.

Flowing into Lake Assunpink, the creek continues its westward course through the management area into Mercer County. Leaving the preserve at the village of Windsor, the Assunpink flows through rural backcountry, parklands, and swamps into the city of Trenton, where it finally debouches into the Delaware River.

The name first appeared as Asinpinck affallit (falls) in Lindeström's 1655 map. The creek was first mentioned by name in the October 10, 1677, deed to land south of the middle and upper reaches "of Sent Pinck Creek at the falls" (New Jersey Archives, Liber B:1–4). It subsequently appeared in something more closely resembling its current form, first as Asunpink in a deed dated June 4, 1687 (ibid., Liber M:447–49), and next in a December 16, 1689, deed to land whose boundary markers passed "over three branches of Assinpinck Creek" (ibid., Liber D:418). A different, much smaller, and long-forgotten "creek called Assinipink" sharing the same etymology was also mentioned flowing into the Hudson River from the east bank of the Hudson Highlands just north of the present-day Bear Mountain Bridge in an Indian deed to land in the area dated July 13, 1683 (Budke 1975a:53–55).

AWOSTING (Litchfield County, Connecticut; Passaic County, New Jersey; and Ulster County, New York). Whritenour thinks it likely that Awosting comes from a Munsee word, *awasahtung*, "the other side of the mountain." In New York, Awosting Lake is located in the heart of the Minnewaska State Park Preserve in the present-day town of Rochester. The name first appeared as a Shawangunk Mountain hill or foothill identified as Aiaskawasting at the northwestern end of the vast tract of land that took in much of northern Orange County, purchased by New York governor Thomas Dongan on September 10, 1684 (Fried 2005:32). Like many place names, Awosting has been moved around a bit. The name itself had been all but forgotten for more than 150 years when entrepreneurs building on the region's growing reputation as a desirable resort destination began to take an interest in its revival. Sometime during the 1850s, a local businessman trading on the romantic associations sparked by Indian names rechristened the blandly named Big Pond in the Shawangunk Mountains as Lake Awasting, a more accessible version of the long-forgotten Aiaskawasting. Newly revived Awosting was soon adopted as the name of the small hamlet and its service road built at the base of the Shawangunk cliffs below the lake. The lake subsequently became the site of a boys' summer camp named Camp Awosting, which was closely associated with the Quaker Mohonk School operated between 1900 and 1942 by Smiley family members who owned the Mohonk and Minnewaska resorts at the northern end of the Shawangunk Ridge. The boys' camp was acquired by new owners, who kept the name Awosting when they moved the facility to its present location in Litchfield County, Connecticut. The Lake Awosting locale in New York subsequently served as a coed camp from 1949 to 1966 and as a retreat site between 1968 and 1970. The lake was finally sold to the state of New York in 1971 in the first parcel of seven thousand acres acquired for conversion into parkland. The name has also been applied to the ninety-foot-high Awosting Falls located several miles north of the lake. It has been further adopted as the name of another hamlet, this one in northern New Jersey along the eastern shore of Greenwood Lake in West Milford Township.

BRAINARDS (Middlesex and Warren counties, New Jersey). The small hamlet of Brainards in Harmony Township is located in New Jersey just across from the place where Pennsylvania's Martins Creek flows into the Delaware River. Brainerd Lake, Dam, and Cemetery are located many miles to the southeast in the village of Cranbury in Middlesex County. Each place is named for two brothers who worked as missionaries among Indians in the region during the mid-1700s. The eldest of these brothers, David Brainierd, lived in Pennsylvania across the Delaware River from the present-day Brainards locale when he worked with the Indians at the Forks of the Delaware from 1743 up to his death in 1747. His brother John accompanied many of David's converts back to their old homes at and around Cranbury in central New Jersey when they were forced to leave the Forks during the 1750s. Both men worked from their cabin near the Indian town of Sakhauwotung on the banks of present-day Martins Creek, named for James Martin, who operated the Martins Creek Ferry across the Delaware River at the stream's mouth during the late 1700s. Employees working on the section of the Belvidere Delaware Railroad operated by the Pennsylvania Railroad later adopted Martins Creek as the name for the station they built on the Jersey side of Martins Creek Ferry. Residents living near the station ultimately adopted Brainards, the slightly revised name of the village's early twentieth-century post office, to distinguish their community from the Martins Creek village on the other side of the river. The Brainerd Cemetery and the Brainerd Lake and Dam carry on the name in Cranbury Township.

BUCKWHA (Carbon and Monroe counties, Pennsylvania). Buckwha is a relocated contraction of *pockhápócka*, a variation of Pohopoco that Heckewelder stated was a Munsee name for the Lehigh Gap. Today, Buckwha is the name of a 12.6-mile-long creek that flows into Aquashicola Creek. Rising just southwest of the Monroe County hamlet of Saylorburg in Ross Township, Buckwha Creek flows through adjoining Eldred Township to its junction with

Aquashicola Creek at the Carbon County community of Little Gap in Lower Towamensing Township some seven miles northeast of the Lehigh Gap. It is possible that an even more altered form of *pockhápócka* may have traveled farther east to Buckabear Pond, one of a cluster of bear-themed place names in West Milford Township, Passaic County, New Jersey, which include Bearfort Mountain and Bear Swamp Lake.

CAHOONZIE (Orange County, New York). Cahoonzie is the name of a hamlet, a lake, and a street in the town of Deerpark. The name was first mentioned as "the old Kehoonge foot-path near the Delaware River" in a note dated May 11, 1804, written by local surveyor Peter E. Gumaer (Ogden 1983:26). A few years later, Gumaer made several references to a locality first identified as Kehoonge and later, in 1853, as Kehoonzie (ibid.:230). The village at the present-day Cahoonzie locale was also called Pleasant Valley by its founders during the early 1790s. No families going by the name of Cahoon or its synonyms Colloquhan or Calhoun appear in Pleasant Valley records. Folk traditions unsupported by documentary evidence hold that the hamlet was named for Cahoonzie, the purported chief of the area's equally imaginary Cahoonshee Indians. A grave containing a skeleton assumed to be his remains was unearthed during roadwork in 1905. The otherwise unidentified body was subsequently ceremoniously re-interred in the local churchyard. The name may also be a somewhat garbled form of Kenoza, the Ojibwa name for "pike" popular in rural resort areas.

The transformation of the spelling of what probably was the Delaware place name Kehoonge, unrelated to the fictional Cahoonshee chief or Indians, to Cahoonzie may preserve a faint trace of the kinds of local sectional passions that helped plunge the nation into civil war in 1861. A town named Calhoun, including lands carved from the eastern part of Deerpark, was set up in 1825. The new town's name honored South Carolina senator John C. Calhoun, who was then widely admired in the area for his support of issues near to the heart of New Yorkers during the War of 1812. Calhoun soon became more closely associated with state nullification of federal laws thought to threaten slavery. Local residents hostile to the idea of nullification arranged to have the town's name changed to Mount Hope in 1833. Adoption of the spelling of Kehoonge that finally produced the present-day form Cahoonzie around this time may represent a more or less subtly veiled local expression (perhaps in the form Calhouns's, a vernacular double plural in the tradition of youse) of continued support for the embattled senator and his politics among voters in Deerpark.

CAKEPOULIN (Hunterdon County, New Jersey). Cakepoulin Creek flows for seven miles through New Jersey Piedmont country, from its headwaters just west of the hamlet of Mount Salem in Alexandria Township, to its junction with the South Branch of the Raritan River two miles below the town of Clinton. The name was first mentioned as "as branch of Raritan River called Capooauling" in the November 11, 1703, deed to a tract of land between the South Branch of the Raritan and the Delaware River (New Jersey Archives, Liber AAA:434–35). Indians may have used the short, level stretch of land between the head of Cakepoulin Creek and an upper reach of Neshisakawick Creek as a portage route when traveling between the Raritan and Delaware valleys. A version of the early spelling of the word has recently been resurrected as the name given to the Capoolong Creek Wildlife Management Area, a sixty-eight-acre preserve managed by the New Jersey Department of Environmental Protection.

CAMPGAW (Bergen County, New Jersey). Campgaw is currently the name of a ridgeline, the mountain at its highest point (732 feet above sea level), a park, and a hamlet at the eastern end of the New Jersey Skylands. The name first appeared in colonial records as Campque, one of five tracts in the area sold by Indians putting their marks on the Ramapough deed dated October 10, 1700 (Budke 1975a:77–78). Colonists moving in shortly afterward kept the name on their maps. Campgaw residents gave the name to the post office opened in their village in 1898. The office closed when the New York, Susquehanna, and Western Railroad's Campgaw Station (also called the Kunschitown Station) took over local postal duties between its construction in 1909 and its closing in 1966 (Wardell 2009:16–17). Today, Campgaw Ridge is a five-mile-long stretch of highlands that forms the northernmost part of the Watchung Mountains. Campgaw Mountain is a focal point for the 1,351-acre County Reservation of the same name located on land managed by the Bergen County Parks Department since 1961.

CANARSIE (Kings County, New York). Many philologists think Canarsie is a Delaware word for "long grass." A cognate of Canarsie occurs in the form Canarese in the state of Delaware. Conoy, a similar-sounding word that Iroquois people used when talking about Piscataway Indians in nearby Maryland, is probably an Algonquian rendering of *ganawagha* (Kenny 1961:5–6). Heckewelder thought the name of the Kanawha River in West Virginia came from a related word. Whatever its etymology, Canarsie has become a byword for Brooklyn. Ask anyone in the borough—most will tell you that the

Canarsees were Brooklyn's original inhabitants. It turns out that they are partially correct. Like its counterpart Minisink far to the west, Canarsie was the last place inhabited by the borough's first people. Dutch purchase of three flats that Indians variously identified as Castuteeuw and Keskataw on the island they called Sewanhacky in 1636 (Gehring 1980:5–6) represents the earliest known record of Indians in present-day Canarsie. The name itself first appeared in something more closely resembling its current form in a January 21, 1647, deed to "a certain tract of land on the south side of Long Island called Canarise" (Gehring 1980:45). English settlers moving to present-day Jamaica in 1656 initially adopted Canarise as the name for their settlement. The Dutch had previously called it Rustdorp, "restful village"; the English briefly called it Crawford before finally settling on Jamaica. English settlers in 1666 also creatively mangled Canarsie into Conarie See (Canarsee Sea), their equally short-lived name for Jamaica Bay (Grumet 1981:5–9). Standardized as Canarsie by the time the local post office adopted the name between 1850 and 1855, it has remained on local maps as one of Brooklyn's most famous neighborhood names up to the present time.

Canarsie has attracted more than its share of folklore. Various writers have claimed that Canarsees were one of the thirteen original tribes of Long Island, that they were the Indians who greeted Hudson in 1609 and sold Manhattan to the Dutch in 1626, that they were allegedly all but exterminated by a Mohawk war party after missing their wampum tribute payment deadline, and that they were the nation whose last survivor's burial shroud was sewn by a local matron sometime around 1790 (Furman 1875:22). None of these legends are substantiated in contemporary records. Extant documents simply state that Canarsie was a place where colonists noted the presence of an Indian settlement between 1647 and 1684.

CANTIAGUE (Nassau County, New York). Whritenour finds that one of Cantiague's defunct colonial-era variants, Ciscascata, resembles a Munsee word, *asiiskuwaskat*, "it is muddy grass." Cantiague Rock is a four-cubic-foot-square boulder that has marked the border between the towns of North Hempstead and Oyster Bay since 1745. It was moved in 1964 to its current location in Cantiague Park (opened in 1961), a 127-acre recreation area managed by the county in Hicksville. Although the story of the rock was probably invented sometime during the 1700s, the word first appeared a century earlier as the boundary marker noted on May 20, 1648, as "a point of trees called by the Indians Ciscascata or Cantiag (Cox 1916–40[1]:625–26). The place called Canteaiug was reaffirmed as a point of trees in the 1653 Oyster Bay "Old Purchase" (ibid.[2]:670–71). In 1698 descendants of the original

Indian deed signers identified a marked white oak at the place where the old point of trees was thought to stand as "the right or true Cantiague" (ibid. [2]:244).

CANTITOE (Westchester County, New York). Cantitoe is a street name in the village and town of Bedford. The name was first noted during the early 1800s as the village of Cantacoc (Gordon 1836:766) and as Cantitoe Corners near the Jay family estate at the present-day junction of Cantitoe and Jay streets near the Katonah's Woods locale. Robert Bolton (1881[1]:3) reported that the otherwise undocumented name was probably a local folk dialect version of Katonah's name, originally given to the nearby woods where a boulder was thought to cover the sachem's grave. Over time this belief evolved into the current tradition holding that boulders in Katonah's Woods mark the graves of Katonah, his wife Cantitoe, and their children.

CAPOUSE. See **KAPPUS**

CATASAUQUA (Lehigh and Northampton counties, Pennsylvania). Heckewelder thought Catasauqua sounded much like a Delaware word, *gattosacqui* or *gattosachgi*, "the earth is thirsty [i.e., it wants rain]." Today, Catasauqua is the name of several municipalities, a creek, and a lake in the Lehigh Valley. The borough of Catasauqua and Catasauqua Lake are in Hanover Township. The village of North Catasauqua, just above the borough, lies in East Allen Township. Rising farther inland in East Allen, Catasauqua Creek flows through North Catasauqua on its way to its junction with the Lehigh River in the borough of Catasauqua. The Lehigh County community of West Catasauqua is located directly across the Lehigh River. Catasauqua Creek was first mentioned as Cattosoque in 1735 (Lambert and Reinhard 1914). The 1770 William Scull map shows a stream identified as Mill Creek at the locale. Another map made in 1848 identifying the run as Calesoque Creek indicates that the original name had already been resurrected when the borough of Catasauqua was formally laid out in 1853. Two years later, the name was given to the borough's first post office. Today, many of the most significant houses in the heart of the old village are preserved in the National Register's Catasauqua Residential Historic District designated in 1984.

CATSKILL (Delaware, Greene, Sullivan, and Ulster counties, New York). Although previous suggested etymologies for this place name first recorded during the early seventeenth century vary from cataract and lacrosse (from the Dutch *kat*, "tennis racket") to *kasteel* (Dutch for "castle) creek, Catskill almost certainly is a Dutch name for cat (perhaps mountain lion or bobcat) creek. Dutch and early English settlers frequently referred to Native people living along and around the Catskill Creek as Catskill Indians. A mostly Mahican-speaking community, the Catskills, like the Taconics on the opposite bank of the Hudson, also included a number of intermarried Esopus, Wappinger, and other Munsee-speaking people.

CAUMSETT (Suffolk County, New York). Caumsett State Historic Park Preserve is located in the village of Lloyds Neck in the town of Huntington. The 1,600-acre preserve is managed by the New York State Office of Parks, Recreation, and Historic Preservation in cooperation with a consortium of local preservation groups. Caumsett Preserve consists of land originally purchased by publishing magnate Marshall Field III for a country estate in 1921. Consulting local records, Field discovered that the Indians who sold Lloyds Neck on September 20, 1654, called it Caumsett (C. Street 1887–89[1]:4–5). He gave the long-forgotten name to his property, which soon became a showplace among the manors on the North Shore. The state of New York retained the name when they acquired Field's estate for a state park in 1961. The Caumsett facility was designated a State Historic Park Preserve in 2010.

CAVEN (Hudson County, New Jersey). Caven Point juts out into New York Harbor behind Liberty Island in present-day Jersey City. It was first noted at its present location in a survey return dated May 12, 1668, as Kewan, a point and tract of upland and meadow (Winfield 1872:56–57). The absence of colonists in the area bearing similar-sounding names such as Quinn, MacKeown, or Cavanaugh indicates that Kewan may be a fragment of an Indian name. It was still on the map in its anglicized form, Caven, when Central Railroad of New Jersey officials ordered laying of track and construction of docks at the sheltered anchorage of Caven Point Cove during the late 1800s. These soon supported the substantial Caven Point Army Depot that still operates at a reduced level as the Caven Point Army Reserve Center. Caven also survives as the name of a nearby road and avenue.

CHAPPAQUA (Westchester County, New York). Whritenour suggests that Chappaqua may be an anglicized equivalent of a Munsee word, *shapakw*, "mountain laurel." Today, Chappaqua is the name of a hamlet in the town of North Castle. At least one document noted by Robert Bolton (1881[1]:361) indicates that a ridgeline in the area was known as the Shappequa Hills during colonial times. Other records show that Quakers, who moved to the area during the 1720s, established a meeting at a place they variously referred to as Shapiqua and Shapequa by 1745. The Chappaqua Friends Meetinghouse built in 1753 is part of the Old Chappaqua Historic District, listed in the National Register of Historic Places in 1974. The name was already being used in its present-day form by the time the Chappequa Mineral Spring Resort was noted at the locale in Gordon's (1836:768) gazetteer.

CHEESECOTE (Rockland County, New York). Whritenour finds that Cheesekokes, an early spelling of the word, sounds similar to a Munsee word, *chiishkohkoosh,* "the robin." Today, Cheesecote Mountain Town Park and nearby Cheesecote Pond are in the town of Haverstraw. Together they are the only surviving reminders of the name mentioned as "a certain tract of land and meadow . . . called Cheescocks" on deeds signed on December 30, 1702, and June 12, 1704 (Budke 1975a:87–88, 92–93). The land taken up by these deeds was combined on March 25, 1707, into the Cheesecocks Patent that snuffed out Indian title to a good chunk of present-day Rockland County. The name may have first been recorded on May 30, 1694, in a receipt for an earlier and since-lost deed for "a certain tract of land lying and being in the county of Orange commonly called and known by the name of Quasspeck" (ibid.:74). Whatever Quasspeck's location or affinities, surveys of land taken up by the Cheesecocks Patent conducted between 1735 and 1749 extended for many miles across the then-contested boundary between the provinces of New York and New Jersey. A map depicting patent boundaries of tracts claimed by colonists in the area drafted by John De Noyelles in 1759 noted Cheesecoke Mountain (presently called Cheesecote Mountain, elevation 971 feet) and identified present-day Minisceongo Creek (see entry herein) as Cheesecocks Brook. Hard feelings caused by the boundary dispute, combined with the potential scatological implications of the last part of the name, led to the name's being all but eradicated from regional maps. Revisionists finally changed the spelling of Cheesecocks Mountain, the last feature to retain the name, to the evidently bowdlerized present-day version of Cheesecote by the turn of the twentieth century.

CHEESEQUAKE (Middlesex County, New Jersey). Nora Thompson Dean thought that Cheesequake sounded very similar to a Southern Unami word, *chiskhake,* "land that has been cleared." Today, Cheesequake Creek, Cheesequake State Park, and the nearby neighborhood of Cheesequake are located in Old Bridge Township. Colonists first noted the existence of a creek variously identified as Cheesquakes and Cheesequaques in an Indian deed to land in the area dated February 25, 1686 (Budke 1975a:65A–65B). The name stuck. Residents used the word Cheesequake when talking about the local three-mile-long tidewater creek (Gordon 1834:120); the large saltwater marsh spreading out across the creek's lower reaches; and the lowlands above the high-water marks of the creek and the marsh. Claybanks at the head of the creek attracted potters to the area during the early 1800s. Inhabitants of the small community of Jacksonville that grew up around the potteries adopted the name Cheesequakes for the post office opened there by 1855. The state of New Jersey began acquiring land near the village for the projected Cheesequake State Park in 1938. Opened in 1940, the park has grown to take in 1,569 acres managed by the New Jersey Division of Parks and Forestry. These lands include many miles of hiking trails and the Cheesequake Natural Area created by the state in 2010 to protect environmentally vulnerable portions of the park.

CHICKEN (Fairfield County, Connecticut). Today, the name of several prominent eighteenth-century sachems named Chickens associated with the Lonetown and Schaghticoke Indian communities (one of whom was known as Sam Mohawk) adorns two large rocks in western Connecticut. Chicken's Rock is a large boulder that sits on the southwest shore of Great Pond in the town of Ridgefield. Farther north, a large outcrop known as Chicken Rock juts into Candlewood Lake. One of the bearers of the name was the brother-in-law of an even more influential sachem farther west along the Hudson River named Taphow (Connecticut State Archives, Indian Series Papers 1:92). Taphow was one of several men from western Connecticut who became prominent sachems along the Highland border between New York and New Jersey during the late 1600s and early 1700s. Chickens's predecessor and namesake Chickheag was a sachem in his own right on the colonial Connecticut–New York border. If he followed Munsee matrilineal postmarital residence rules requiring that husbands move in with their wives' families, Chickens would have moved east to Connecticut, away from his and his sister's family home in New Jersey following his marriage to a woman belonging to his brother-in-law Taphow's family. It was there that he may have been given or assumed his predecessor's name. Another descendant bearing the name Chickens exchanged his people's last hundred acres at Lonetown for two hundred acres farther north at Schaghticoke in 1748 (Wojciechowski 1985:112). Members of the Chicks family living at the Stockbridge-Munsee Reservation in Wisconsin may trace descent from Chickheag and Chickens.

CHINGARORA (Monmouth County, New Jersey). Whritenour thinks that a Northern Unami word, *chiingaaluwees*, "stiff tail," sounds much like Zinckkarowes, the earliest known variant of the name recorded in the first Indian deed to land in the area signed on August 5, 1650 (New Jersey Archives, Liber

1:6–7). Today, Chingarora Creek is a small tidewater stream that flows north from its source in Hazlet Township to form the border with the Aberdeen Township borough of Keyport. It later appeared as Changororissa in the October 17, 1664, authorization (Christoph and Christoph 1982:53) for the June 5, 1665, purchase of land around Changarora (Municipal Archives of the City of New York, Gravesend Town Records, Deeds:74). Clammers harvested particularly succulent Chingarora oysters prized by Gilded Age gourmands in New York City from the oyster beds lying just offshore from the mouth of Chingarora Creek until pollution caused the Raritan Bay shellfishery to collapse in the first quarter of the twentieth century. Although the fame of Chingarora oysters has since faded into forgetfulness, the name remains on modern-day maps as Chingarora Creek and as Chingarora Street in Keyport.

COBAMONG (Westchester County, New York). Cobamong Pond is located in the town of Bedford. The name appears twice in local colonial records (Marshall et al. 1962–78[2]:132, 161), first as the "land and meadow of Cohamong" in a May 2, 1683, Indian deed, and again under the same spelling in a deed dated September 6, 1700. The toponym has adorned Cobamong Pond for at least the past 150 years (French 1860:703).

COCHECTON (Sullivan County, New York, and Wayne County, Pennsylvania). Heckewelder thought Cochecton sounded much like *gischiéchton*, Delaware for "finished, completed." Today, Cochecton is the name of several municipalities clustered around a stretch of level lowlands in the Upper Delaware River Valley. John Reading, Jr. (1915:98), first noted Cochecton as "the Indian town called Kasheton" that he visited on June 30, 1719. Reading was a member of the first surveying party sent by the governors of New Jersey and New York to locate the Station Point marking the westernmost boundary of the contested border between their provinces. His sketch maps located a number of Indian towns scattered around what the party agreed was the right spot (on the Pennsylvania side of the river about two miles north of present-day Skinners Falls).

An Orange County militia officer named Thomas De Kay was sent to what he called Cashighton in December 1745 at the height of King George's War. He was ordered to find out the truth behind a rumor that Indians at the locale were getting ready to support a forthcoming French and Indian attack. Arriving at the town, he found between ninety and one hundred men and their families who had fled from their homes farther east in Orange and Ulster counties for fear of an English attack. Finding out that the rumors were false, he invited the Indians to a peace conference in the spring. They subsequently came to the Orange County seat of Goshen, where they put their marks to a treaty with New York and signed deeds finally surrendering title to the vast expanse of Hardenbergh Patent lands in the Catskills that they had contested since 1707 (Ruttenber 1906b).

Forced to give up their homes on the New York side of the Delaware River, the Indians moved to the far bank along a narrow stretch of flat land in present-day Damascus, Pennsylvania. Sachems from that town, whose numbers included displaced Forks Indian leaders Nutimus, Teedyuscung, and Kappus, were among the Indians who later sold all of their remaining lands in Pennsylvania north of the Delaware Water Gap to New Englanders in 1754 and 1755 as many Delawares inspired by early French and Indian victories over the British during the runup to the Seven Years' War in America finally decided to go to war against their erstwhile neighbors.

No Indians were living at the place New Englanders called Cushietunk when Connecticut settlers began moving into the region in 1760, after peace returned to the area following the signing of treaties at Easton, Pennsylvania, in 1758. Cushietunk became a springboard for New England penetration into the province throughout the Pennamite-Yankee Wars that sputtered on for the next forty years. The name remained on Pennsylvania maps in part as evidence buttressing New England claims to the region. The hills behind the town were identified as Cushichton Mountain in the 1770 William Scull map of the Pennsylvania frontier.

New Yorkers began settling around a place they called Cushetunk across the river shortly after the end of the Revolutionary War. People flooded into and through the area following completion of the Cochecton-Newburgh Turnpike in 1810. The hamlet's maturation into a village was signaled by the opening of a post office bearing the name of Cochecton in 1811. The community that developed around the post office became the nucleus of the town of Cochecton established in 1828. In 1853 Cochecton became a center for railroad business on the Delaware, Lackawanna, and Western line (later called the Erie Lackawanna Railroad). Two other hamlets in the town, East Cochecton and Cochecton Center, were also established around this time. Today, the National Register of Historic Places includes the Old Cochecton Cemetery (designated in 1992); the Cochecton Presbyterian Church (1992) and Railroad Station (2005); and the Cochecton Center Methodist Episcopal Church (2000). New York State Routes 17K and 17M follow much of the Cochecton-Newburgh Turnpike's old right-of-way. A continuation of the road west of the river (Pennsylvania State Route 371) is still called the Cochecton Turnpike. A local road running between Narrowsburg and Yulan named Cochecton Turnpike Road in honor of the old thoroughfare does not follow any part of its namesake's original route. Also surviving is the massive brick-walled ruin of the Cochecton Pumping Station built by the Standard Oil Company in 1879. The station housed eight oil-fired pumps that propelled crude oil along the company's pipeline from its oil fields in western Pennsylvania to its refineries at Bayonne, New Jersey, until 1924.

COCKENOE (Fairfield County, Connecticut). Cockenoe Island is the name of one of the Norwalk Islands that lie offshore from Norwalk and Westport on Long Island Sound. Like several other islands in the chain, Cockenoe Island is separated from the mainland by Cockenoe Harbor. Both the island and the harbor are named for Cockinsecko, a prominent Indian culture broker in the region (Tooker 1896) who, identified as Conkuskenoe, put his mark alongside those of Winnapauke and several other sachems on the February 15, 1651, Indian deed to the lands along the shores of Cockenoe Harbor (Hurd 1881:483–84). Local residents evidently adopted the spelling of Cockenoe, popularized by William Tooker's biography (1896) of this notable Indian intermediary. Cockenoe Island is currently owned by the town of Westport and maintained as a bird sanctuary.

COMPO (Fairfield County, Connecticut). Today, Compo Beach, Point, Cove, and Road in the town of Westport bear the name of Compow, the sachem of Indians living east of Saugatuck noted by Norwalk colonists in 1660 (Selleck 1886:74–75). The present-day locale was variously identified as Compo and Compaw Neck in Indian deeds signed between 1661 and 1680 (in Wojciechowski 1985:96, 98, 103). Developers selling lots for their Compo Beach summer resort on the shores of Cockenoe Harbor resurrected the name during the early decades of the twentieth century. The resort's exceptional architectural and design values helped secure the place's listing in the National Register of Historic Places as part of the Compo-Owenoke Historic District in 1991.

CONASKONK (Monmouth County, New Jersey). Nora Thompson Dean suggested that Conaskonk meant "place of tall grass," basing her translation on a Southern Unami word, *kwënàskung*. Whritenour provides a Northern Unami equivalent consisting of *gun*, "long or tall," *askw*, "grass," and the locative suffix *onk*. Today, Conaskonk Point juts into Raritan Bay in the village of Union Beach in Hazlet Township. The name Conneskonck was first mentioned as one of the tracts of land just south of the bay sold by local Indians on August 5, 1650 (New Jersey Archives, Liber 1:6–7). It has been identified on nautical maps as Conaskonk Point Shoal since at least 1898. The place came to widespread attention when local preservationists rallied to prevent construction of an oil tank farm at Conaskonk Point during the 1960s and 1970s. The Conaskonk Point wetland is now regarded as a prime migratory bird observation site.

COPIAGUE (Suffolk County, New York). Copiague is a hamlet in the southwestern part of the town of Babylon. The area was the scene of a lengthy colonial land dispute between Oyster Bay town settlers who claimed that the lands belonged to their Massapequa Indian clients and Huntington townsfolk who purchased the lands from the Montaukett Indians. The name itself first appeared in colonial records as a neck "commonly called by the Indians Coppiage" in a March 8, 1666, settler's deed to land in the area (C. Street 1887–89[1]:84). The name recurred as Copyag Neck in Indian deeds dated November 11, 1693, and July 2, 1696 (in ibid.[2]:121–23, 189–91). The present-day Great Neck River was identified as the Copiag River in the latter deed. Copiague hamlet residents variously called their community South Huntington, Great Neck, and Amityville South before deciding on its present name in 1895.

COPLAY (Lehigh County, Pennsylvania). This place name is a garbled colonial rendering of a Shawnee personal name. Coplay borough is named for the youngest son of Shawnee chief Paxinosa. His name was variously spelled Kolapechka, Kolapeka, and Copelin in colonial records. Thought to have been born somewhere near his namesake community sometime around 1721, he was later known as a warrior who presumably died fighting against the British on the Ohio frontier during the French and Indian War (Greene and Schutz 2008:211). Colonists settling in the vicinity of present-day Coplay after 1740 initially named the place Schriebers, before changing it to Lehigh Valley. Local residents there evidently resurrected the name Coplay from colonial records when they split off from Whitehall Township to form a borough of their own in 1869. They have since named the borough's municipal recreation area Kolapechka Park. Another aspect of local community life linked with its Indian name is preserved in the Coplay Cement Company Kilns listed in the National Register of Historic Places in 1980.

CORLEAR (New York County, New York). Present-day Corlear's Hook Park is located on a stretch of landfill along the banks of the East River on Manhattan's Lower East Side. The original mainland was the location of Nechtank, an overnight camp catering to Indians visiting nearby New Amsterdam operated by a local sachem named Numerus. It became one of two places where Lower Hudson River Indians (the other was Pavonia, in present-day Jersey City) taking refuge under Dutch protection from a Mohawk or Mahican raid were murderously assaulted by detachments ordered out by Governor Kieft on the night of February 25–26, 1643. At least 120 people, mostly women, children, and elders, were killed in the attacks, which led to the bloodiest phase of Governor Kieft's War against the Indians living around New Amsterdam. Iroquois and River Indians subsequently addressed English governors of New York ceremonially as Corlaer to honor the memory of respected frontier go-between Arendt van Corlaer, a bearer of the family name who drowned in Lake Champlain while on a diplomatic mission to Canada in 1667.

COXING (Ulster County, New York). Whritenour finds that Coxing sounds very much like a Munsee word, **kooksung*, "place of leaf-eating worms, i.e., caterpillars." Today, Coxing Kill is an eleven-mile-long Shawangunk Mountain stream that begins at Lake Minnewaska. From there it flows north through the Mohonk Preserve to its junction with Rondout Creek just west of the village of Rosendale. Coxing Kill was first mentioned in an April 19, 1700, Indian deed to a tract of land called Kochsinck on the west bank of Rondout Creek (New York State Library, Indorsed Land Papers [2]:276). Builders have highly regarded Coxing stone as a superior raw material for millstones, paving stone, and similar products since quarries began operating around Rosendale during the early 1700s.

CRECONOOF. See **CRICKER**

CRICKER (Fairfield County, Connecticut). The one-mile-long Cricker Brook flows from the Hemlock Reservoir into the Samp Mortar Reservoir on Mill Creek in the town of Fairfield. The name first appeared in Fairfield town meeting minutes apportioning planting commons "above Crecroes Brook, upon the neck there," on January 3, 1661 (in Schenk 1889:105). It more recently began to be mentioned in geological reports published during the 1890s. Cricker is probably a much-anglicized spelling of Cockenoe, locally spelled as Creconoes, Crehero, and Crecono in deeds signed at Fairfield between 1679 and 1686 (in Wojciechowski 1985:95–110). It also presently adorns Creconoof Road in the Lake Hills development.

CROSSWICKS (Burlington, Monmouth, and Mercer counties, New Jersey; also in Warren County, Ohio, and Montgomery counties, Pennsylvania). Philologists currently find little value in one of my favorite etymologies, a still-popular example dating back to at least 1844. In the version published in Becker (1964:15), Crossweeksung is identified as a Delaware word meaning "place of separation," not just as that formed by a fork in a river, but also as a reference to a menstrual separation hut presumably located at the site. Whatever its etymology, the name Crosswicks adorns several places in central New Jersey in the general area where Lindeström noted a stream named Packaquimensj Sippus in his 1655 map. Today, Crosswicks is best known as the name of a twenty-five-mile-long stream that flows north and east from its headwaters in the Burlington County Pinelands at Fort Dix and McGuire Air Force Base. From there it meanders into Monmouth County before becoming the border separating Burlington and Mercer counties. The creek finally flows into the Delaware River at Bordentown. West Jersey Quakers started their first settlement at the present site of the village of Crosswicks shortly after Indians sold much of their land between the Assunpink Creek and Rancocas River on October 10, 1677 (New Jersey Archives, Liber B:4). Initially naming the settlement Chesterfield, they changed it to Crosswicks by the time they built their first meetinghouse in the village in 1693. The name Chesterfield survived, adopted by another village south of Crosswicks and given to the surrounding town when it incorporated in 1798. The name Crosswicks first came to the attention of local colonists in places identified as the "small river called Croswicksum" and as the Indian town of Crosswicks on its banks, mentioned in a deed to land in the area dated January 14, 1686 (Salter and Beekman 1887:250). Crosswicks Creek became a well-known stream to settlers who repeatedly referred to it as a boundary line in subsequent land purchases between 1689 and 1709. The place later became a Presbyterian Indian mission town around 1745–46; still later, Indians from southern New Jersey concluded a separate peace agreement there with the provincial government on January 9, 1756 (ms. on file, New Jersey State Library). At this treaty they agreed to exchange all of their remaining lands for a reservation. According to terms were agreed upon at the Easton Treaty on February 23, 1758, the Indians gave up claims to territory below a line running from the Raritan River westward to Lamington and on to the Delaware Water Gap and accepted a reservation in the Pinelands at Edgepillock soon named Brotherton in the area now known as Indian Mills (New Jersey Archives, Liber I-2:245–47). Today, in New Jersey, in addition to serving as the name of the creek and village, Crosswicks is also a common street, club, and corporation name, as well as the name of the 1,500-acre greenway maintained by the Monmouth County Park System along the section of creek that flows through the county. Use of this place name extends beyond New Jersey to the hamlet of Crosswicks just east of Jenkintown, Pennsylvania, and farther west to Crosswick, a village in the greater Cincinnati area, first platted in 1821.

CROTON (Hunterdon County, New Jersey; Bronx and Westchester counties, New York). The name Croton first appeared in colonial records as a stream called the Scroton River in the October 19, 1696, Indian deed to land in the area abstracted in Robert Bolton (1881[1]:362–63). It was again mentioned as Scrotons River in an August 4, 1705, sale document (New York State Library, Indorsed Land Papers [4]:58). It was subsequently thought to be an Indian name for the present-day Croton River. Local legends claim that Croton was the name of an otherwise undocumented sachem. Engineers designing and building massive stone structures for the Croton Reservoir, Dam, and Aqueduct during the early nineteenth century may have more closely associated the word with monumental Greek and Roman constructions in southern Italy at Crotone (Crotona in Latin). Whether the name is Munsee, Greek, or Latin, it was given to the village at the river's mouth christened Croton-on-Hudson and later traveled to central New Jersey, where it has graced the small hamlet of Croton west of Flemington since the mid-1800s. It subsequently reached the Bronx in the form of Crotona Park by the early 1900s.

CUSHETUNK (Hunterdon County, New Jersey). Whritenour proposes that an early form of the name Coshawson sounds somewhat similar to a Munsee word, **kaanzhasun*, "great or amazing rock." Today, Cushetunk is the name of a mountain, a reservoir, a park preserve, and several roads in and around Clinton Township. Water pumped into the bowl-like depression at the center of Cushetunk Mountain from the nearby South Branch of the Raritan River fills the Round Valley Reservoir, constructed in 1960 by building dams across gaps along Cushetunk's almost circular rim. The place first became a prominent landmark for colonists when it was mentioned as the "mountain called by the Indians Coshawson" in an Indian deed to land in the area dated November 11, 1703 (New Jersey Archives, Liber AAA:434–35). Cushetunk was later adopted as the name of the railroad station and the hamlet that grew up around it just to the east of the mountain in neighboring Readington Township. More recently, the Hunterdon County Department of Parks and Recreation has given the name to its 380-acre Cushetunk Mountain Nature Preserve.

CUTLASS (Morris County, New Jersey). Whritenour suggests that Cutlass resembles a modern-day spelling of a Jersey Dutch word, *katelos*, "wildcat or bobcat," that served as the source of the nickname they gave to a local Indian whose town was noted in survey returns as Cutlosses Plantation on August 12, 1753, and Catloss Plantation on December 9, 1755 (New Jersey Archives, East Jersey Survey Book S3:353 and S4:65). Today, the name adorns parts of an old road running between Butler and Riverdale that is broken into sections by New Jersey State Route 23 and Interstate 287. Its westernmost part is known as Cutless Road in Butler; it is known as Cutlass in Kinnelon Township and as Cotluss at its eastern end in the borough of Riverdale.

CUTTALOSSA (Bucks County, Pennsylvania). Cuttalossa is the name of a creek, a road, and an unincorporated hamlet that was first mentioned in a 1702 survey return for lands laid out in present-day Solebury Township "at Quaticlassy" (W. Davis 1876:296n.). Remote, secluded, and too small to support heavy industry, the four-mile-long Cuttalossa Creek Valley escaped major development. The stone mills built along its banks during the eighteenth and nineteenth centuries seem to have only added to the scenic valley's rustic charm. Local residents hope that the Cuttalossa Valley Historic District's designation in the National Register of Historic Places in 2002 will help maintain the qualities that have long drawn poets such as John Greenleaf Whittier and artists such as the New Hope School Bucks County painter Daniel Garber to the locale.

DANSKAMMER (Orange County, New York). Danskammer is one of those non-Indian words that refer to an Indian place, in this case a Dutch name for "dance chamber," which is a fair description of the flat sheet of rock colonists noted as a place where Indians were seen holding celebrations. Although the name is often attributed to Henry Hudson, Fried (2005:116) shows that Dutch colonist David Petersz de Vries made the first recorded reference to the place in a journal entry noting what he said were riotous Indian celebrations at what he called the Danskammer on April 26, 1640. Fried further notes that another Dutchman writing nearly a quarter of a century later during the Esopus War observed that "Indians [at Danskammer] made a great uproar every night, firing guns and kintekaying [a Delaware trade jargon word for dancing]." Luis Moses Gomez built his fieldstone mill house, the oldest extant Jewish residence in North America (listed in the National Register of Historic Places in 1973), near the Danskammer between 1710 and 1714. His property was ultimately purchased by a wealthy local farmer named Edward Armstrong who built his Armstrong Mansion at Danskammer Point in 1834. It stood there until it was acquired in 1874 by the owner of the local brickworks who subsequently demolished the building to make room for further expansion of his production facilities. Another neighbor salvaged and reused the mansion's five enormous granite Ionic columns. Today, these stand together in splendid isolation in the sculpture gardens of the Storm King Art Center. The nearby Danskammer Point Light was built in the 1880s and demolished sometime after it was decommissioned during the 1920s. The present-day town of Newburgh's Danskammer Point and the Danskammer Generating Station, built there in 1951, preserve the name in the area.

DELAWARE is a ubiquitous place name on regional maps. Universally regarded as the general name of the region's Indians, Delaware is not an Indian word, as mentioned in the introduction. It is instead an English word, the name of Thomas West, baron de la Warr, second governor of Virginia, who sent the ship that carried the men who gave his name to Delaware Bay in 1610. Only later did it spread from the bay and its river to become a name identifying the Indian people who lived along its banks. Although the name Delaware has been on regional maps far longer than many other toponyms brought back to life by resurrectionists, examples tend to be generic references not associated with historically documented places or events specifically mentioning Delaware Indians or communities. It stands alone as the original proper name of the Delaware River, Delaware Bay, and the state of Delaware. Delaware counties, townships, parks, lakes, streets, schools, and avenues can be found throughout the region and across the continent.

DOCK WATCH (Somerset County, New Jersey). Whritenour proposes that Dock Watch resembles an anglicization of a Delaware word, *takwahchuw*, "short hill." Today, Dock Watch Hollow, Brook, and Road pass through a gap in the Second Watchung Mountain that forms the boundary between Bridgewater and Warren townships. Dock Watch has long been thought to be a much-altered spelling of the name of William Dockwra, a prominent London polymath, founder of the penny post, and an East Jersey proprietor whose agents purchased land in the valley. Modern-day Dock Watch first appeared as Doct wache Brook in an Indian deed to land in the area signed on July 26, 1708 (New Jersey Archives, Liber K-large:131). It was next mentioned as Doquateches Hollow in a November 10, 1714, Indian deed to an adjacent tract (Special Collections, Alexander Library, Rutgers University, Middlesex County Early Records, Land Deeds, 1714–22:234–36). Both deeds were signed several years before Dockwra, an absentee owner who never visited the province, purchased an interest in lands there (Siegel 1989).

EQUINUNK (Wayne County, Pennsylvania). Heckewelder thought that Equinunk sounded exactly like a Delaware word meaning "place where we were provided with articles of clothing, where wearing apparel was distributed to us." The Equinunk Creek watershed is an extensive Upper Delaware Valley stream system. Fifteen-mile-long Equinunk Creek forms the boundary between Manchester and Buckingham townships as it flows toward its junction with the Delaware River at the village of Equinunk. Little Equinunk Creek is an adjoining but separate stream of nearly equal length that joins the Delaware some miles south of the village of Equinunk at a place called Stalker. Pennsylvania proprietary officials authorized the erection of the 2,222-acre Safe Harbor Manor at the mouth of Equinunk Creek in 1721 (Goodrich 1880:99). The area remained an untenanted tract mostly frequented by hunters and passing log rafters even after a Bucks County Quaker named Samuel Preston purchased the land for his Equinunk Manor in 1791. Several families subsequently bought land, built houses, and erected saw and tanning mills at Equinunk during the early 1830s. Those that survive from this era are included in the Equinunk Historic District listed in the National Register of Historic Places in 1999.

ESOPUS (Greene and Ulster counties, New York). Goddard (1928:237 and PC: June 14, 2012) thinks Esopus is a pidgin word whose Munsee form "was apparently *só·psi·w* or *wsó·psi·w*, 'person from *sópəs*.'" Basing his etymology on an analysis of Sypous, an early orthography of Esopus evidently recorded by someone who spoke Dutch (the language has no palatalized *s* like the English *sh*), Whritenour thinks it likely that the name comes from a Munsee word, *shiipoosh*, "little river." Words sharing similar etymologies include Sepoose, the now-defunct name of a small stream mentioned in an August 24, 1674, Indian land sale in Monmouth County (New Jersey Archives, Liber 1:271[68]–270[69]) and, less apparently, as Paramp Seapus in an Indian deed dated May 9, 1710 (ibid., Liber I:317–19).

The Esopus Creek rises at Lake Winnisook below the northeast slope of Slide Mountain in Catskill State Park. Geological studies indicate that the bowl-like depression at the headwaters of the creek was formed by a meteor that struck the area 375 million years ago. Today, the Esopus Creek generally flows west around the outer rim of this ancient crater into the Ashokan Reservoir. On its way to Ashokan its volume is increased by water pumped through the Shandaken Aqueduct from the Schoharie Reservoir eighteen miles farther north.

Those waters not carried to New York City by the Catskill Aqueduct pass through the main Ashokan Dam sluice gates into the deep rocky gorge below the reservoir barrage. From there the creek meanders across the broad flats where the Late Woodland ancestors of the historic Esopus Indians located many of their settlements. Passing the city of Kingston, the Esopus turns north paralleling the Hudson River for ten miles to the place where it falls into the Hudson River at Saugerties.

Archaeological evidence excavated from sites along the creek around Hurley and Kingston corroborates colonial accounts of intensive Indian settlement along the lower course of the creek first noted in a Dutch map drawn in 1616. Violent encounters between Indians and settlers in the area between 1658 and 1664, and the Nicolls Treaty made in 1665 (named for English governor Richard Nicolls who presided over the treaty after forcing the Dutch to surrender their New Netherland colony) helped secure a place for the Esopus nation in the memories of successive generations of mid–Hudson Valley residents. Less well remembered are the many land sales, meetings, the annual renewals of peace and friendship mandated by the Nicolls Treaty (Scott and Baker 1953), and the other get-togethers that put Esopus Indians into close contact with colonists in and around Ulster County for more than a century after the Esopus War. Most Esopus Indians continued to support the crown in accordance to what they regarded as their Nicolls Treaty obligations when the Revolutionary War broke out in 1775. Colonists responded by forcing the Indians to abandon their homes. Moving first to nearby Oquaga, just west of the Catskills, they later moved into the massive loyalist Indian refugee camps at Niagara after American militiamen burned Oquaga and the other Upper Susquehanna Indian towns in 1778. Most moved to reserves set aside for them by British authorities in present-day Ontario after the end of the war, where one descendant was later known by the nickname só·pši·w (Speck and Moses 1945:2, 17).

People living near Kingston adopted the name of the creek for their new town of Esopus when they broke away from the city in 1811. The decision to retain Esopus on local maps was made at the height of the classical revival that saw New York communities adopt names such as Rome, Syracuse, and Utica. Esopus's similarity to Aesopus, the Greek author of wisdom literature, doubtless helped keep the name on the maps. Since then, Esopus has been applied to nearly every conceivable place capable of bearing a name in the area.

FORT NECK (Nassau County, New York). An English word for an Indian place. Today, Fort Neck is a neighborhood in the village of Seaford at the southwest corner of the town of Oyster Bay. Archaeological remains, including patterns of soil stains preserving evidence of a ditch and postholes of a fortified palisade wall, have fueled folklore traditions identifying the place as the site of a Mohawk attack and as Massepe or one of the other forts on western Long Island attacked by colonial troops in 1644 during Governor Kieft's War. While colonial records are silent on these subjects, deeds still on file in local town archives show that Massapequa Indians living at Fort Neck sold their last lands there between 1686 and 1697 (Grumet 1995). The suburban residential Harbor Green development built during the late 1930s now takes up most of the land at Fort Neck above the high-tide line. The Fort Massapeag Archaeological Site, located in a community park on a small piece of land originally set aside for its preservation, was designated a National Historic Landmark in 1993.

GOFFLE (Passaic County, New Jersey). Although Goffle is often regarded as an Indian name, it is in fact a Dutch word for "fork" that colonists used when referring to a split in the Indian trail that passed through the place.

Today, Goffle is the name of a brook, a park, and a nearby mountain range. The Goffle Mountain ridge (formerly called Totoway Mountain) is a part of the First Watchung Mountain that runs north for several miles above the west side of the city of Paterson. Goffle Brook and Goffle Brook Park (a landscaped recreation area listed in the National Register of Historic Places in 2002) are located in Paterson's Hawthorne neighborhood.

GOLDEN HILL (Fairfield County, Connecticut). The Golden Hill Reservation is located on a quarter acre of land set aside by the state of Connecticut in 1876 as a campground for Paugusset tribe members visiting or working in the nearby community of Trumbull. The state established a second 106-acre reservation for the tribe far to the east of their homeland in the town of Colchester in 1979. Golden Hill has been formally recognized as an Indian reservation by Connecticut authorities since 1659.

GOWANUS (Kings County, New York). Today, Gowanus is the name of a Brooklyn neighborhood, creek, canal, expressway, and bay. The name first appeared as a tract of land called Gouwanes in a deed dated April 5, 1642 (Gehring 1980:13–14). Subsequent colonial chroniclers variously referred to the stream as the Gouwanisse Kill and Gouwanus Creek. The name's resemblance to Gouwe, a swampy river in the Netherlands, probably accounts for settlers' early acceptance of the Indian name. Persistent Dutch colonists continued using the name even after English settlers moving into the area after 1664 increasingly called the place Mill Creek. The nineteenth-century romantic revival of interest in Indian place names kept Gowanus on the maps just as its use waned to a historic minimum following the end of the Revolutionary War. Many local residents subsequently took pride in associating themselves with the mythical Indian chief alleged to have borne the name. Rival genealogists fought back, claiming it was originally named for the Huguenot immigrant Pieter Gouane. Extant records showing that Gouane arrived in New York more than forty years after Gowanus made its first appearance support its status as an Indian word. The place called Gowanus, meanwhile, festered into a cesspit of industrial pollution. Construction of the expressway named for the creek during the 1950s and more recent efforts to clean up the much-besmirched waters flowing below its massive pillars have kept the name Gowanus on the minds of city officials, corporate executives, conservationists, and copy editors who see to it that news of their latest run-ins reaches their readers.

HACKENSACK (Bergen County, New Jersey; Dutchess and Rockland counties, New York). Heckewelder thought that Hackensack sounded very much like *hackinksáquik*, a Delaware word he said meant "the stream which discharges itself into another on low level ground, that which unites itself with other water almost imperceptibly." Nora Thompson Dean's translation was "place of sharp ground," from a Southern Unami word, *ahkinkèshaki*, a word very similar to Achkinckeshaky, an early recorded variant of the name. Today, Hackensack is the name of the river, the meadowlands, the city, and the narrow valley above it that lies just beyond the Palisades on the west bank of the Hudson River. The Hackensack River rises as a series of small streams just below the High Tor at South Mountain on the southeast edge of the Hudson Highlands in Rockland County. Dammed to create reservoirs for local communities at several points along its course, the stream flows south into Bergen County past the city of Hackensack across the Meadowlands to its junction with the Passaic River at the head of Newark Bay.

The efforts of Hackensack sachem Oratam to restore peace during the Indian wars that ravaged the region kept both names in the news between 1641 and 1664. Colonists began penetrating Hackensack country in 1668 only after the Elizabethtown and Newark purchases opened the way. One of the settlers who bought land in the area promptly named it New Barbadoes. Like the name Jamaica, in Queens County, New York, New Barbadoes celebrated the settler's lucrative trade links with a major Caribbean island entrepôt.

The Hackensack Indians themselves had to move elsewhere following the sale of their last lands in their home territories during the 1690s. Hackensack remained on the maps as the name of the river and as that of a township located between the Hackensack and Hudson rivers from 1693 to 1871. Residents of neighboring New Barbadoes Township changed their community's name to the city of Hackensack in 1921. Today, the name Hackensack occurs widely in and around the river valley. The name also spread beyond its original borders fairly early on. Residents of a little crossroads village south of Poughkeepsie in Dutchess County, New York, for example, named their post office New Hackensack when it opened in 1836 (Kaiser 1965).

HAKIHOKAKE (Hunterdon County, New Jersey). Today, Hakihokake Creek's headwaters rise along the south-facing slopes of Musconetcong Mountain in Holland Township. They then join together just below the hamlet of Little York to fall into the Delaware River in the borough of Milford. Settlers moving into the area during the mid-1750s called the creek and the mill erected at the junction of its upper branches Quequacommissicong (Moreau 1957:21). Modern-day standardized spellings of the name have occurred in local records since 1834 (Gordon 1834:154).

HARIHOKAKE (Hunterdon County, New Jersey). Harihokake Creek is a small stream in Alexandria Township that flows from Musconetcong Mountain into the Delaware River just south of the mouth of its near namesake Hakihokake. The nearly identical spellings of both creeks may represent a kind of modest corrective of an earlier undocumented surveyor's error mistaking the adjoining separate drainages as parts of a single watershed.

HARSIMUS (Hudson County, New Jersey). Harsimus Cove is located on the banks of the Hudson River in Jersey City. The name first appeared in Dutch patroon (manor lord) Michiel Paauw's July 12, 1630, Indian deed to Hoboken as a place called Ahasimus south of the place identified in the document as "the land called by us Hobocanhackingh" (Gehring 1980:1). Paauw managed to acquire the former tract, identified this time as Harsimus, in another deed signed a few months later (ibid.:3–4). The name remained in local use even after colonial officials formally renamed the area Bergen in 1660. Harsimus presently is regarded as the oldest municipality in New Jersey. The Harsimus locale became an important railhead and ferry port during the nineteenth century. Reclaimed marshes fronting on the Hudson River became known as Harsimus Cove during this period of development. The Passaic and Harsimus rail line originally established on these lands by the now-defunct Pennsylvania Railroad continues to operate as a freight carrier. Well-preserved examples of the many residences built alongside the area's rail yards and warehouses were designated as the Harsimus Cove Historic District, listed in the National Register of Historic Places in 1987.

HAVERSTRAW (Rockland County, New York). A Dutch word for "oat straw," also spelled Averstraw and Heardstroo in deeds dating to the 1680s, Haverstraw was used to identify the Indians whose settlements were concentrated within a triangular stretch of lowland flanked by the Hudson Highlands to the north, the Ramapo Mountains on the west, and the South and Hook mountains to the south. Today, it is the name of a town, a city, and of Haverstraw Bay, a wide section of the Hudson River above the Tappan Zee. Dutch colonists used the name to identify Indians also referred to as Rumachenancks, Reweghnoncks, and Rechgawawancks (the latter thought by Whritenour to sound like a Munsee word, *leekuwaawunge*, "sandy hill") living in and around present-day Haverstraw from the 1640s to the end of the seventeenth century (Grumet 1981:25–26). A longtime center of the brick-making industry, Haverstraw has been depicted on local maps at its present location since colonial times. The place was also among the first municipalities to be erected as a town in the Hudson Valley. The local post office adopted the name when it opened in

1815 (Kaiser 1965). The postmaster of the office opened at North Haverstraw in 1834 changed its name to Grassy Point in 1847. North Haverstraw finally disappeared from maps altogether when it was incorporated into the present-day town of Stony Point formed in 1865. The West Haverstraw community founded during the early 1800s remains a village in the town of Haverstraw.

HOBOKEN (Hudson County, New Jersey). Heckewelder was perhaps the first writer to note that Hoboken sounded much like a Delaware word, *hopokan*, "a tobacco pipe." Whritenour thinks early mention of Hoboken Hackingh in colonial records probably represents a Northern Unami–based Delaware pidgin combination of the Munsee word *hopoakan*, "tobacco pipe," with *hakink*, "at the land of." Ruttenber thought it was a conflation of a Dutch folk etymology *hoebuck*, "high hill" (the root of the name of the village of Hoboken, on the Scheldt River below Antwerp in present-day Belgium) with a rendering of a Dutch word for "hook" (a word that also resembled the Delaware pidgin locative ending *hackingh*), to produce a conflated Dutch folk etymology, "High Hill Hook." Hoboken first appeared in local records in two deeds signed in 1630, the first chronicling Michiel Paauw's purchase of "the land called by us Hobocanhackingh" on July 12 and the second his November 22 purchase of land nearby (both in Gehring 1980:1–4). Dutch colonists initially called the locale Pavonia, a Latin form of Paauw. The area subsequently was called Bergen until 1820, when it was incorporated as the city of Jersey City. Local residents never stopped using one spelling or another of Hoboken when writing about the area. Entrepreneur John Stevens adopted the name when he purchased the land at the north end of Bergen in 1804. The ferry and short line he operated at Hoboken helped bring visitors to his resort at the locale. His Hoboken Land and Improvement Company began attracting more permanent residents soon after its founding in 1838. The growing village of Hoboken was included within the newly established county of Hudson that broke off from Bergen County in 1840. By 1849 the population had risen sufficiently to warrant erection of Hoboken as a township. Six years later, Hoboken became a city. Today, the Stevens Institute of Technology occupies the Hook of Hoboken that juts into the Hudson River at the north end of the city.

HOCKHOCKSON (Monmouth County, New Jersey). Whritenour (in Boyd 2005:436) suggests equally possible translations like those for the Munsee word Ho-Ho-Kus followed by the locative suffix *ung*, "place of." Today, Hockhockson Brook, Swamp, and Park are located in the hamlet of Tinton Falls in Colts Neck Township. The name was first mentioned as "a boggy meadow called by the Indians Hochoceung" in a deed to land in the area signed on August 24, 1674 (New Jersey Archives, Liber 1:271[68]–270[69]).

HO-HO-KUS (Bergen County, New Jersey). Whritenour thinks that Ho-Ho-Kus sounds similar to two Munsee words: *mehokhokwus*, "red cedar," and *hakhakwus*, "little bottle gourd." Today, Ho-Ho-Kus is the name of a borough and the brook that flows through it. The name first appeared in colonial records as Hochaos Brook in a deed confirmation dated August 8, 1696 (Whitehead et al. 1880–1931[21]:247). One of three known versions of another deed to land in the area dated November 18, 1709, mentions a place where a creek called "Raikghawaik (otherwise Anhokus creek)" falls into the present-day Saddle River (New Jersey Historical Society, Manuscript Group 567, folio 20). Land near the place noted as Hockakens was entered as the birthplace of Marytje Hopper on the September 28, 1718, record of her baptism in the Dutch Reformed Church of Hackensack. Later documents note that the locale, subsequently called Hoppertown, was renamed New Prospect. Ho-Ho-Kus has followed a tortuous route to its present position as a borough name. Residents living along Hohokus Brook at Hoppertown continued to also call their community Hohokus (see Gordon 1834:158). Things started getting complicated when people living many miles north at the upper end of Hohokus Brook adopted the name for their own township in 1849. Ten years later, the postmaster at New Prospect adopted the village's other name for his newly opened Ho-Ho-Kus post office. In 1886 the southern part of Hohokus Township joined with the Ho-Ho-Kus postal district to form Orvil Township, named for a prominent local resident. The Ho-Ho-Kus post office again changed its name when residents established the borough of Orvil there in 1905. Confusion almost became total when Orvil residents decided to rename their borough Ho-Ho-Kus in 1908. The situation was finally relieved in 1944 when the people of Hohokus Township, by then much diminished in size by multiple removals of several other communities to neighboring jurisdictions, agreed to change its name to Mahwah.

HOKENDAUQUA (Lehigh County, Pennsylvania). Heckewelder thought Hokendauqua sounded something like *hackundòchwe*, a Delaware word meaning "they are searching for land." Today, Hokendauqua is the name of a creek and the municipality named after it located half a mile south of the stream's confluence with the Lehigh River in Whitehall Township. Runners hired by the Penn family to establish the extent of the line marking the Walking Purchase spent the night of September 19, 1737, on the banks of the creek. The name first appeared in colonial records twenty years later in what may be an ironic reference to the event when one of the hired walkers and the surveyor Nicholas Scull sent out by the Penns recalled that the Indians at the place they identified as Hockyondocquay had not hidden their anger over the walk. Present-day Hokendauqua was first laid out in 1855 just as work was being completed on the Catasauqua and Fogelsville feeder connecting the area with the Lehigh Valley Railroad main line. It soon became a company town dominated by the Thomas Iron Works. The works closed when the valley's iron industry collapsed during the 1960s. Today, Hokendauqua is a mostly residential community.

HOMINY (Monmouth County, New Jersey). The Hominy Hills are part of the low ridge of upland that forms the divide between the inner and outer coastal plains at the Earle United States Naval Ammunition Depot. The Hominy Hill Golf Course is located just north of the depot in Colts Neck Township. The Hominy part of the name is apparently a folk contraction of what Boyd (2005:436) identified as the old name Homhomonany, translated by James Rementer in Boyd's study as a Southern Unami word for "herring stream or river," from *haamo*, "herring," or *haames*, "pilchard," and *hanneck*, "stream." Folk renderings of this Delaware word into the form used to identify ground parched cornmeal commonly called hominy grits (originally a Virginian Algonquian word, *uskatahomen, usketchamun*, "that which is treated" [i.e., ground or beaten; in Bright 2004:170] given to other places across the nation) doubtless eased its transition onto New Jersey maps.

HONK (Ulster County, New York). Today, Honk is the name of a hill, a waterfall, and a lake located above the falls in the Rondout River Valley just west of the city of Ellenville. Water from the lake cascades over falls that gradually descend in stages seventy feet through a narrow gorge into the fast-running stream that flows past Ellenville before turning north to parallel the Shawangunk Ridge. Although dams have been built to harness Honk's power, the cascade barely visible through the trees lining its banks most times of the year has retained much of its natural beauty. Colonists first noted the place as "a certain fall in the Rondout Creek called by the Indians Honckh" northwest of "land called Nepenach" in an Indian deed to the area signed on March 22, 1707 (New York State Library, Indorsed Land Papers [4]:92). Local residents apparently later inserted Honk into the middle of the current standardized spelling of Kerhonkson located six miles north of the falls. The altogether separate Falls of the Neversink River was noted on July 6, 1705, in "the County of Ulster on the Waggackemek [Neversink] Creek or River beginning by a certain tract of land called by the Indian name of Nawesinck from thence running down said creek to a certain fall in said creek called Hoonckh" (Ulster County Records, Deed Book AA:353). This appearance of the same name in two different locales in the same general area suggests that it may be either a local term for waterfall or a mistaken repetition.

HOPATCONG (Morris and Sussex counties, New Jersey). Whritenour suggests that Hopatcong sounds similar to a Munsee word, *(eenda) xwupeekahk*, "where there is deep water or a lot of water." Today, Hopatcong is the name of New Jersey's largest lake as well as the borough and state park located along its southernmost shores. It was first noticed in the Mackseta Cohunge (thought by Whritenour to sound like the Munsees' word *meexksiit takwaxung*, "place of the red turtle") purchase of August 13, 1709, as "a run of water called Hapakonoesson where the same comes out of the mountain" (New Jersey Archives, Liber I:210–11). Surveyor John Reading, Jr. (1915:41), next noted the name in its more current form and function as "a large lake at the head of one of the branches of Muskonethcong, called Huppakong" on May 20, 1715. Settlers called the lower lake Great Pond and its upper reach Little Pond, and they gave the name Brookland to the village that grew up alongside the lower lake. Residents changed its name to Brooklyn, a word doubtless more familiar to the workers who began construction on the dam that joined the Great and Little ponds together in 1831. The enlarged lake served as a feeder pond supplying water for the Morris Canal built to carry coal from the Lehigh Valley across northern New Jersey to the port of New York. Construction of the Central Railroad of New Jersey line paralleling the canal route opened the region to tourism during the 1850s. The Lake Hopatcong Yacht Club listed in the National Register of Historic Places in 1997 marks just one aspect of the many summer recreational activities developed at the locale. Growth of the year-round population led to the incorporation of Brooklyn as a borough in 1898. Realizing that people were primarily drawn to the area by the lake, borough residents adopted the by then well known name of Hopatcong to adorn for their community two years later.

HOUSATONIC (Fairfield and Litchfield counties, Connecticut; Dutchess County, New York). Trumbull (1881) presented and endorsed fluent Schaghticoke language speaker Eunice Mahwee's etymology for Housatonic: *Hous'atenuc*, "over the mountain." The upper-middle part of the Housatonic River Valley bends westward across the northeasternmost edge of the Greater New York area where Fairfield, Litchfield, and Dutchess counties meet along the southern foothills of the Berkshire Mountains. Schaghticoke, the Indian town at Kent, Connecticut, that became a major Native mission and refugee community during the early 1700s, is the Housatonic Valley locale most notably associated with the Indians of Greater New York. Several Indians associated with the northern New Jersey sachem Taphow became active in land affairs in this Connecticut valley during the first quarter of the eighteenth century. Descendants of the Esopus squaw sachem Mamanuchqua moved to the Moravian Pachgatgoch mission established at Schaghticoke during the century's second quarter. Daniel Nimham and other members of his Wappinger community subsequently made Schaghticoke a stop along their far-flung network that extended from western Massachusetts to central Pennsylvania between the 1740s and the 1770s.

INDIAN FIELD (Bronx County, New York). Indian Field is one of the very few places bearing the name Indian that is clearly linked to an event involving Native people in the region during the colonial era. Today, Indian Field is located in the Katonah section of the Bronx in Van Cortlandt Park on East 233rd Street between Van Cortlandt Park East and Jerome Avenue. The locale marks the site of the Revolutionary War Battle of Cortlandt Ridge fought on August 31, 1778. Outmaneuvered by British dragoons, Captain Abraham Nimham, commander of the Continental Army's Stockbridge Indian Rifle Company, his father Daniel Nimham, and at least fifteen other men from the unit were killed in the engagement. Abraham and Daniel Nimham and several other Wappingers in their unit were direct descendants of ancestors who had lived in and around the Bronx long before colonists or redcoats ever came to America.

INDIAN LADDER (Warren County, New Jersey). Indian Ladder Cliff is located at the Delaware Water Gap in Knowlton Township. It first appeared in colonial records as an Indian ladder (probably a notched tree trunk) that John Reading, Jr. (1915:41), and his party propped up on May 19, 1715, to clamber over a twenty-foot-high rock blocking their path at a round hill he called Penungauchongh (present-day Manunka Chunk). Although Reading noted other Indian ladders elsewhere during his surveys, the name came to be fixed to a rock formation near Manunka Chunk resembling the kind of improvised tree ladder used by the surveyor's party.

INDIAN WALK (Bucks County, Pennsylvania). The area in the village of Wrightstown just north of the place where colonists and Indians began their walk to determine the extent of land to be taken up by the Walking Purchase on September 19, 1737, is often locally referred to as Indian Walk. The name has also been recently applied to a development and its access road a few miles farther east in Mechanicsville.

IRESICK (Middlesex County, New Jersey). Probably a Northern Unami name for a sachem also spelled Irooseeke and Jakkursoe. Heckewelder thought the name came from a Delaware word, *achôkoêt*, "one who takes care of a thing, a preserver," a reference, Heckewelder believed, to a wampum keeper. Today, the three-mile-long Iresick Brook named after the sachem flows through Old Bridge Township into Duhernal Lake on the South River. The stream's name was first recorded as Irasaca's Brook in a deed confirmation to land in the area dated October 7, 1700 (Whitehead et al. 1880–1931[21]:311). The brook's namesake, who still lived in the area in 1700, was the nephew of the sachem Ockanickon whom the latter designated as his successor on his deathbed in 1682 (Budd 1685). Other documents show that this leader, who was noted in one document (Stilwell 1903–32[33]:449) as a father of Weequehela, signed many deeds to lands in and around the area from 1676 to 1701.

JOGEE (Orange County, New York). A Dutch nickname for a Munsee man. Today, Jogee Road and Joe Gee Hill are located in the hamlet of Slate Hill in the town of Minisink. Both names mark the memory of Joghem, one of the principal Indian signatories of the deed that conveyed much of present-day Orange County to New York governor Thomas Dongan on September 10, 1684 (Budke 1975a:56–59). Colonists knew him better as Jochem, a Native leader noted in a number of other lands in the area between 1663 and 1709 under such various spellings of his Indian name as Sherikam, Choukass, Sekomeek, and Shawaghkommin.

KAKIAT (Rockland County, New York). Today, Kakiat County Park is located in the hamlet of Montebello in the town of Ramapo. Most examples of this place name, which in various forms, is attached to hills in northern New Jersey, eastern Pennsylvania, and Ulster and Westchester counties in New York, come from the Dutch word *kijkuit*, "lookout." Kakiat Park in Rockland, however, comes from a Delaware place name, Kakyachteweeke, mentioned in the June 25, 1696, deed that transferred Native title to much land in present-day Rockland County to colonists (Durie 1970). Later colonists knew the tract as the Kakiat Patent.

KAPPOCK (Bronx County, New York). Today's Kappock Street in Riverdale is a fragment of Schorakkapoch, the Indian name for the Cold Spring Harbor section of Spuyten Duyvil Creek mentioned in an October 8, 1666, patent to land first purchased in 1646 by Adriaen van der Donck (in Robert Bolton 1881[2]:585). Van der Donck was also known as the *jonkheer*, Dutch for "young lord," memorialized today as the name of the city of Yonkers just across the city line in Westchester County.

KAPPUS (Hunterdon County, New Jersey; Lackawanna and Wayne counties, Pennsylvania). Another Dutch nickname for a Munsee man. Today, Kappus Road in Alexandria Township, New Jersey, perpetuates an old German family name (Kappus is German for "cabbage") that also became a nickname of the influential Munsee sachem Tammekapi. In the account of his life given in 1749 to the Moravian minister who baptized him and gave him yet another new name, Salomo, Tammekapi said that he was born around 1672 in present-day Rocky Hill, New Jersey (Moravian Archives, Indian Mission Records, Box 313, Folder 1, Item 3). Colonists spelled his name as Tameckapa, Toweghkapi, and Tawakwhekon on deeds to lands in and around central and northern New Jersey signed between 1694 and 1744. Several of these deeds listed his nickname in the forms Cappos, Capoose, and Capohon. Known to the Moravians as the king of the Delawares at the Forks, he moved to the Lackawanna Valley after being forced from his lands in the Lehigh Valley following the Walking Purchase of 1737. His village near present-day Scranton, called Capoose's Town by settlers, became one of the main Munsee townsites in the area until the outbreak of the French and Indian War forced its abandonment in 1755. A Moravian scribe noted that Salomo died a year later farther upriver at the Indian town of Tioga (Moravian Archives, Baptism Register 133). Colonists moving into his old home on the Lackawanna River originally adopted his name for their settlement. They changed it to Providence, its current name, in 1770. A since-drained reservoir at the locale was also named for the sachem. Today, Scranton's Capouse Avenue, Capouse Mountain in Lackawanna County's Scott Township, the Wayne County hamlet of Callapoose in Sterling Township, and Kappus Road in Alexandria preserve a map memory of this New Jersey Indian man who rose to become a Munsee sachem in Pennsylvania.

KATONAH (Fairfield County, Connecticut; Bronx and Westchester counties, New York). Today, Katonah is the name of a hamlet and lake in the town of Bedford in Westchester County, a street in the town of Ridgefield, Connecticut, and a neighborhood in the North Bronx. All are named for the local leader who signed most of the land deeds transferring Indian title to colonists in and around Bedford between 1680 and 1708. Residents there informally gave the sachem's name to local features ranging from Katonahs Woods north of the town center to Katonahs Meadow at its south end (Duncombe et al. 1961:66). The local post office adopted the name sometime between 1850 and 1855. The hamlet of Katonah made its first formal appearance on Westchester County maps in 1897 when the stream-level village of Whitlocksville was relocated and given its present name at its higher elevation above the floodplain inundated by the waters impounded by the Cross River Reservoir. The heart of the relocated hamlet was listed in the National Register of Historic Places in 1983. Local developers purchased land several miles east of the village and dammed the small pond on the property for a development they named Lake Katonah in 1926. Farther east in the town of Ridgefield, Connecticut, local businessmen earlier built a meetinghouse they dubbed Catoonah Hall in 1860. Today, Ridgefield's Catoonah Street remains the only cartographic reminder of the now-vanished hall's existence. Another entrepreneur gave the name Katonah to his development built during the early 1900s just east of the place where the somewhat similar-sounding Croton Aqueduct enters the borough of the Bronx.

KAWAMEEH (Union County, New Jersey). Kawameeh Park is located in Union Township. The park was built by the Union County Park Commission in 1940. The name they adopted for the park exactly reproduces one of the spellings used by a colonial scribe to identify Cowescomen, one of the principal Indian signatories who put their marks onto the Elizabethtown Deed that conveyed most of Union County to colonists on October 24, 1664 (New Jersey Archives, Liber 1:1). Cowescomen later became more familiar to local settlers as Queramack, a sachem who participated in many land sales in and around the Raritan Valley between 1664 and 1684.

KENSICO (Westchester County, New York). Another place named for the Indian culture broker Cockinseko, who signed a deed to land in the area on November 22, 1683 (in Robert Bolton 1881[2]:536). Today, the Kensico Reservoir and Dam are located where the towns of North Castle, Mount Pleasant, and Harrison meet. In 1848 Bolton (ibid.[1]:468) noted that the hamlet of Kenseco, formerly called Robbins Mills, was originally located just below Rye Pond along the upper reaches of the Bronx River. The Kensico post office opened at the locale one year later (Kaiser 1965). In 1885 the people who operated Kensico's mills and post office moved to Wright's Mills, located where the first Kensico dam held back water destined for New York City. They stayed together until waters that backed up behind the significantly larger present-day dam completed in 1917 forced them to move again. This time community residents scattered to other places less likely to be flooded. Their defunct community's name continued to appeal to New Yorkers, who retained it for the newly enlarged dam and reservoir. Kensico Reservoir continues to be a major collecting point for New York City–bound water carried by aqueducts from the Delaware and Catskill reservoir systems. The Westchester County

Department of Parks, Recreation, and Conservation manages the formally landscaped Kensico Dam Plaza at the foot of the dam.

KEOFFERAM (Fairfield County, Connecticut). Keofferam Road is located on the shores of Greenwich Cove on Old Greenwich Neck. It was originally designed to be the main street of the Keofferam Park development built at the locale during the 1910s. The developers borrowed the name from local history books mentioning Keofferam as one of the Indian signatories to the July 18, 1640, deed to land in the area (Hurd 1881:365–66).

KERHONKSON (Ulster County, New York). Today, Kerhonkson is the name of a creek, a reservoir, and a hamlet in the town of Rochester. Rising in the low hills at the foot of the Shawangunk Ridge several miles behind the hamlet, Kerhonkson Creek flows into the community reservoir and on to its junction with the Rondout Creek a few miles farther on. The name was first mentioned in the minutes of the October 7, 1665, Nicolls Treaty as Kahankson, a creek marking the southwestern boundary of land Indians surrendered to the colonists (O'Callaghan and Fernow 1853–87[13]:399–402). Fried (1975:73–77) has surveyed evidence suggesting that the Esopus Indians may have built on the creek a fortified settlement called the First or Old Esopus Fort destroyed by Dutch troops in 1663. The stream next appeared as an upper branch of the Rondout Creek called Kahakasink in the minutes of a treaty renewal meeting that took place on April 27, 1677 (Christoph, Christoph, and Gehring 1989–91(2):57–59). The locale was more firmly located near its current location as a sixty-acre tract called "Kahakasins, being the first land at said Kahankisins," in an April 18, 1683, Indian deed (Ulster County Records, Kingston Town Records [I]:43–44). Fried (2005:41–42) identified a currently unnamed stream one mile north of present-day Kerhonkson Creek as the Kahakasins of 1683. Delaware and Hudson Canal workers gave the name Middleport to the village they built at present-day Kerhonkson after completing the towpath section run through the area in 1831. Fried (ibid.:42) shows that residents resurrected the place's long-forgotten Indian name just a few years before French (1860:668) listed it in his gazetteer. Like other communities along the railroad line that replaced the canal, Kerhonkson transformed itself into a service community for the many resorts that began opening nearby at the turn of the last century. Although the local tourism business has not entirely disappeared, most present-day residents of Kerhonkson commute to workplaces farther from their homes.

KINDERKAMACK (Bergen County, New Jersey). Many writers regard Kinderkamack as a Dutch word having something to do with children at play. Majority opinion among philologists, however, leans toward an Indian etymology of some sort. Whritenour suggests a possible Munsee derivation from *kundakamik*, "praying grounds." Today, Kinderkamack Road follows the course of an old wagon road more or less followed by a ten-mile stretch of Bergen County Route 503. Traced from south to north, the road runs alongside the Hackensack River from River Edge to Oradell. Veering away from the river, it runs north through the borough of Emerson to Westwood, where it crosses the Pascack Brook. From there it runs through the Pascack Valley north to the New York–New Jersey state line at Montvale. The name first appeared in a license dated May 30, 1684, permitting purchase of "two hundred acres of land of the Indians at Kinderkamacke at Hackinsacke above the Mill" (Anonymous 1872:109). Residents who first named the area Old Bridge gave Kinderkamack to their post office twice, first between 1831 and 1844 and finally from 1870 to 1877, when they established a borough they named Etna. Borough residents adopted its present name, Emerson, in 1909 (Wardell 2009:55). At present, Kinderkamack survives as a street name in the area.

KISCO (Westchester County, New York). Whritenour thinks Cisqua and other early spellings of the Indian part of this place name sound like a Munsee word, *asiiskuw*, "mud." Mount Kisco is currently the name of a town at the headwaters of the Saw Mill River. The name first appeared in colonial records as Cisqua, the name of a meadow and river mentioned in the September 6, 1700, Indian deed to land in the area (Marshall et al. 1962–78[2]:161). The local postmaster used the spelling Mount Kisko when he opened a post office at the village recently grown up along the rail line, which had reached the area in 1850 (Kaiser 1965). The name took on its current spelling around the time when Mount Kisco was formally incorporated as a village in 1875. The village became a freestanding town in 1978. The Mount Kisco Municipal Complex at the town center was listed in the National Register of Historic Places in 1997. The name also adorns the three-mile-long Kisco River, which flows through the town into the nearby Croton Reservoir.

KITCHAWAN (Westchester County, New York). Whritenour thinks Kichtawanck, the spelling of Kitchawan thought to most closely approximate the original sound of the name, closely resembles a Munsee word, **kihtaawunge*, "big hill." The modern anglicized form of the name is presently fixed onto two places in Westchester. The best known of these is the 208-acre Kitchawan Preserve managed by the Westchester County Department of Parks, Recreation, and Conservation. Located on the south shore of the Croton Reservoir in the town of Yorktown, it was purchased from the Brooklyn

Botanical Garden, which had operated the property as a research facility for many years. Kitchawan Pond, on the other side of the county, was created in the early 1900s as a focal point for a camp and cottage resort built by Benn Bryon, a physician turned developer from Ridgefield, Connecticut. The spellings of these names indicate that both were directly drawn from Robert Bolton's history (1881[1]:35). Colonists used variants of a name they spelled Kichtawanck when they began documenting relations with Indians living in and around the present-day Croton Valley during the war years of the 1640s. Although many writers, including me, have identified Kichtawanck as an Indian name for the Croton River, specific reference to a stream "called Ketchawan and called by the Indians Sint Sinck" mentioned in the August 25, 1685, deed to land in the area (ibid.[2]:2–3) suggests that, like the sizable Wiechquaesgeck nation and the diminutive Wickers Creek, Kichtawanck people may have identified themselves with the more minor stream today known as Sing Sing Creek located just south of the much larger Croton River. Whatever river they were associated with, Kichtawancks devastated by frontier warfare and epidemic disease sold their last small tracts of land in Westchester and moved to safer ground farther from the colonial settlements by the turn of the eighteenth century. Those trying to return to their old homeland most closely associated themselves with the Wappinger community led by members of the Nimham family. Nearly all of these people ultimately joined Wappingers who moved between the Indian mission at Stockbridge, Massachusetts, and the Indian towns far to the west beyond the Catskill Mountains along the upper branches of the Susquehanna River.

KITTATINY (Sussex and Warren counties, New Jersey; Monroe and Northampton counties, Pennsylvania). Nora Thompson Dean identified Kittatiny as a slightly anglicized version of a Southern Unami word, *kitahtëne*, "big mountain." Whritenour thinks a Munsee version would sound more like **kihtahtune*. The word *Kehtuhtin*, "big mountain," appears in a nineteenth-century Munsee hymnbook (Wampum 1886). Today, Kittatiny is the name of a mountain chain and numerous nearby places on both sides of the ridge. The Kittatiny Ridge is part of the great Appalachian mountain chain called the Shawangunks farther north in New York and the Blue Mountains farther west in Pennsylvania. The name Kittatiny has also been applied to two parts of the Blue Mountains in Pennsylvania: the northern portion between the Delaware Water Gap and Wind Gap along the border of Monroe and Northampton counties, and a smaller stretch farther to the southwest, well beyond Greater New York in Franklin County. The normally taciturn surveyor John Reading, Jr. (1915:93), was so impressed by the height of the ridge at the Delaware Water Gap that he launched into a flurry of adjectives characterizing it as "a very stupendous high hill difficult to be crossed, unless at certain places where runs of water facilitates the passage." The name first appears as the Kekkachtanin Hills in the deed Pennsylvania authorities had the Forks Delawares sign on October 11, 1736, to confirm the Walking Purchase. Lewis Evans noted the location of the Kittatinny Mountains in his 1738 map of the area taken up by the walk (in Gipson 1939). Ten years later, Pennsylvania colonists noted that Delaware Indians referred to the ridge running between the Delaware and Susquehanna rivers as the Kittochtinny Hills (State of Pennsylvania 1851–53(5):407). After the Revolutionary War, people in New Jersey fell into the habit of giving the name Kittatiny to the stretch of the Blue Hills towering above the eastern banks of the Delaware River between the Water Gap and the New York line. Today, the name occurs widely in and around the mountains, adorning everything from the ridge to state parks, lakes, camps, and canoe liveries.

KOLAPECHKA. See **COPLAY**

LACKAWACK (Ulster County, New York). Lackawack sounds like a truncated version of *lechauwêksink*, a Delaware word Heckewelder translated as "the forks of the road." Other names with similar etymologies located elsewhere in Greater New York include Raccawackhacca, mentioned in a December 1, 1684, deed to land at the Forks of the Raritan River (Whitehead et al. 1880–1931[21]:84), the Rockaway River in New Jersey, and the place names Lehigh and Lackawaxen. Today, Lackawack is a small hamlet located in the town of Wawarsing several miles downstream from the Merriman Dam holding back the waters of the Rondout Reservoir. It is the single survivor of the three communities forced to abandon their low-lying locations in the part of the Rondout River Valley flooded by reservoir waters after construction on the dam began in 1937. Lackawack first appeared in colonial documents as Ragawaak, one of six tracts located within the bounds of lands in the town of Rochester mentioned in the Beekman deed dated November 2, 1708 (O'Callaghan 1864:89–90). Local residents also used the name Lackawack as a general term when talking about the upper reaches of the Rondout Creek above Honk Falls. It came to wider notice in the region after provincial militiamen built a fort there during the Revolutionary War. By 1835 the village had grown large enough to warrant the opening of a post office bearing the name at the locale (Kaiser 1965). Both village and post office had to move when the dam planned as early as 1909 for New York City's projected Lackawack Reservoir finally materialized into today's Rondout Reservoir thirty years later.

LACKAWAXEN (Sullivan County, New York, and Pike County, Pennsylvania). Like Lackawack, Heckewelder thought Lackawaxen came from a Delaware word, *lechauwêksink,* "the forks of the road, or the parting of the roads; where the roads take off in various directions." Basing his etymology on its earliest recorded form, Lechawachsein, Whritenour thinks it more likely that it was originally a Munsee word, *(eenda) lxawahkwsiing,* "where there is a forked tree ... where there are forked trees." Today, the name Lackawaxen is most closely identified with a thirty-two-mile-long river that begins as outflow from the Wallenpaupack Dam. The Wallenpaupack Branch of the Lackawaxen River flowed unhindered above Hawley before final enlargement of a dam built there created Lake Wallenpaupack and completely flooded its lower reaches in 1926. The present-day Lackawaxen River flows past the village of Hawley, where it is joined by the waters of its twenty-two-mile-long West Branch. The river runs alongside the Delaware and Hudson Canal through Lackawaxen Township between Honesdale and the place where the stream flows into the Delaware River at the village of Lackawaxen just north of Minisink Ford. The name first appeared in colonial records as a creek called Lechawachsein in the August 21, 1749, deed to lands above the Walking Purchase line sold by a group of Iroquois, Delaware, and Shawnee claimants (State of Pennsylvania 1851–53[5]:407). The stream subsequently was noted as Lechawaxin Creek in the 1770 William Scull map and as Lexawacsein on the 1792 Reading Howell map. The Lackawaxen Valley became an important corridor traversed by boats and barges carrying anthracite to market during the canal era. Today, the Lackawaxen Aqueduct that transported Delaware and Hudson canal boats over the Delaware River above the Minisink Ford is a National Historic Landmark preserved by the National Park Service within the Upper Delaware National Scenic and Recreational River. A local wood road called Lackawaxen Road is located in the town of Tusten in Sullivan County, New York, several miles north of the aqueduct. The name is also preserved in somewhat modified form as Lacawac Sanctuary, a 545-acre preserved Adirondack-style great camp built in 1903 near the shores of Lake Wallenpaupack. The sanctuary's Lake Lacawac is a National Natural Landmark. Both places were listed in the National Register of Historic Places in 1979.

LAHASKA (Bucks County, Pennsylvania). Whritenour thinks Laoskeek, an older recorded form of the name, sounds much like a Northern Unami word, **lawaskeek,* "the middle of the swamp." Today, Lahaska is the name of a creek, a hill, and a hamlet most widely known as a popular market village and tourist destination. The village draws its name from Lahaska Creek, one of the uppermost branches of the Neshaminy Creek whose headwaters rise just west of the village. The name is also applied to a nearby upland frequently called Buckingham Hill. Local tradition holds that the village of Lahaska was known as Hentown during the early nineteenth century (MacReynolds 1976:217). Lahaska Creek and its lower course, today called Mill Creek, were known as Randalls Creek and Randalls Run, respectively, during the early nineteenth century. Lahaska began to appear on local maps following the 1842 publication of a newspaper article recalling how the father of local resident John Watson mentioned a place he called Laoskeek in a poem written in 1805 (MacReynolds 1976:218). Another poem penned forty years later spelled the name Lahasaka (ibid.:219). Ten years farther on, Lahaska finally appeared in its present form as the name of the local village post office.

LAHAWAY (Monmouth and Ocean counties, New Jersey). Whritenour (in Boyd 2005:44) thinks that Lahaway most closely sounds like *lechauwi*, "forked," a Northern Unami cognate of a similar-sounding Munsee word. Today, Lahaway Creek is a seven-mile-long stream that rises below the Lahaway Plantation Dam at Switlik Lake in Jackson Township in Ocean County. Passing over the dam's spillway, the waters of the creek flow into Prospertown Lake and through Upper Freehold Township in Monmouth County. From there Lahaway Creek flows south and west past Lahaway Hill across the lower part of the township to its junction with Crosswicks Creek. The name first appeared in colonial records as the Lechwake tract sold in an Indian deed to land in the area dated April 20, 1699 (New Jersey Archives, Liber H:219–20).

LAMINGTON (Hunterdon, Morris, and Somerset counties, New Jersey). Lamington is an anglicized spelling for "the place called Allamotunk which is where the said river has a considerable fall betwixt two hills" mentioned in an October 13, 1709, Indian deed to land in the area (New Jersey Archives, Liber B-2:274–75). Whritenour suggests this earliest spelling of the word closely sounds like a Munsee word, **alaamahtung*, "underneath or bottom of the mountain." Transformation of the name proceeded from Allamotunk to Lamaconick in 1738, Lamoertonk in 1751, and Lamenton in 1765 (Backes 1919:250). Today, Lamington is the name of a river, a falls, a hamlet, and a street. The nineteen-mile-long Lamington River is a tributary of the North Branch of the Raritan River. The twelve-mile-long upper section of the Lamington in Morris County is also known as the Black River. It rises in Mine Hill Township and flows past the Roxbury Township villages of Kenvil and Succasunna into the more than three thousand–acre expanse of marshlands and woodlands protected within the state of New Jersey's Black River Wildlife Management Area. The river then passes through a succession of narrow, steep gorges, mostly on lands within Hacklebarney State Park, to the long series of cascades today called Potters or Pottersville Falls. Formerly called Lamington Falls, its position in the place where the borders of Hunterdon, Morris, and Somerset counties converge is a reminder of its function as a prominent boundary marker during colonial times. The stream becomes known as the Lamington River as it enters the mostly level stretch of Piedmont valley beyond the falls between Pottersville and its confluence with the North Branch of the Raritan at Burnt Mills. On the way it passes by the hamlet of Lamington whose most significant early buildings are preserved within the Lamington Historic District, listed in the National Register of Historic Places in 1984.

LEHIGH (Carbon, Lehigh, Luzerne, Monroe, and Northampton counties, Pennsylvania). Heckewelder wrote that Lehigh reminded him of Delaware words associated with *lĕcháwâk*, "fork," such as *lechau-hanne*, "forks of streams," and *lechau-wichen*, "the forks or parting of roads or where they meet together," which the Delawares used when talking about the Forks of the Delaware in the Lehigh River Valley. Today, the Lehigh Valley watershed extends across a 1,360-square-mile area on lands straddling the north and south slopes of the Blue Mountain ridge. Its dominant feature is the 103-mile-long Lehigh River. The river was known as the West Branch of the Delaware and the region around it as the Forks of Delaware during colonial times. Two other streams bore the name Lehigh before it was affixed to the present-day river. The first was Lehiethes Creek, a stream initially shown on the Evans map of 1738, whose name was later changed to Brodhead Creek. The second, the Lehigh Creek shown on the 1770 William Scull map and now called the Little Lehigh Creek, is a twelve-mile-long brook that flows into the Lehigh River at Allentown.

The name Lehigh itself came late to colonial records. It was first mentioned as Lechay in 1697, the same year that peace returned to the region following the end of the first French and Indian War that had ravaged the region since 1689. Settlers were drawn to the Lehigh's broad fertile valley only after the second French and Indian War ended in 1714. Among these settlers were Jersey Indians who moved to the Forks after their leader Weequehela was found guilty of murdering a neighbor and hanged in 1727. Lehigh country subsequently became a hotbed of proprietary land shark shenanigans, culminating in the Walking Purchase of 1737 that forced all of the Forks Indians to abandon the region a few years later. Raids launched by the uprooted Indians from the beginning of the final French and Indian War in 1755 to the end of the Revolutionary War in 1782 limited colonial expansion into and beyond the valley. The subsequent discovery of anthracite coal (ideal for fueling steam-powered machines) at the upper end of the Lehigh Valley transformed life in the area. Coal carried by boats running along the Lehigh Canal mostly completed by 1829 powered industrial development in Philadelphia and New York. The valley itself became, at one time or another, a major center for the cement, iron-making, steel, and transportation industries. The dominance of industry in the region was most forcefully reflected by the Lehigh Coal and Navigation Company's status as the only private corporation allowed to own a major waterway in the United States between 1821 and 1966. The region's industrial power made the Lehigh Valley Railroad and Lehigh University among the most widely known corporations bearing a Munsee name. The name also appears elsewhere in the valley as Lehighton, a borough in Carbon County; Lechauweki Springs Park in the city of Bethlehem; and the Lehigh Canal and the Old Lehigh County Courthouse, listed in the National Register of Historic Places in 1980 and 1981, respectively.

LOANTAKA (Morris County, New Jersey). Today, Loantaka is the name of a brook that rises in the city of Morristown. The stream flows south into Loantaka Pond in the 570-acre Loantaka Brook Reservation, managed since 1957 by the Morris County Park Commission. Passing beneath a bridge carrying Loantaka Way across the brook, the stream makes its way to its junction with Great Brook, a headwater of the Passaic River, in the Great Swamp National Wildlife Refuge. The name Loantaka first appeared in a deed signed on March 14, 1745, by descendants of the original signatories to replace an earlier conveyance. It occurred as Loantique, the name of one of three Indians identified as kings whose number also included Quichtoe (also known as Quish, leader of the Indian community at Crosswicks) and Tishewakamin (Tishcohan, a leader of Indians at the Forks of the Delaware who in 1745 tried to fend off eviction from lands taken by the Walking Purchase eight years earlier). Colonists asked the Indians for this confirmation of their rights to land between the first range of the Watchung Mountains and the Passaic River just a week after they claimed that the original unrecorded deed from 1702 was destroyed in a house fire (New Jersey Archives, Liber E-3:105–107). Residents evidently resurrected Loantique's name from the colonial records during the nineteenth century and applied it in its present form to the brook and other features that currently bear the name.

LOCKATONG (Hunterdon County, New Jersey). Whritenour states that Lockatong sounds like a Munsee word, *lokatink*, "place of wheat meal," though he is not sure why Indians not known to have raised the grain in the area would have given the place that name. Lockatong Creek is a sixteen-mile-long stream that rises near the village of Quakertown in Franklin Township and flows through the townships of Kingwood and Delaware to its junction with the Delaware River midway between Byram and Prallsville. A 583-acre state wildlife management area along its banks also bears its name. The first appearance of the name occurred in a November 11, 1703, Indian deed as "a certain brook called Lockatony" identified elsewhere in the document as Lackatong (New Jersey Archives, Liber AAA:434–35). Inclusion of a stream labeled Looking Creek at Lockatong's current location on the 1770 William Scull map shows that at least some settlers had already anglicized the name during colonial times. Locktown, a hamlet along the banks of Wickecheoke Creek in Delaware Township, just one mile east of an upper fork of Lockatong Creek, may be another altered version of the name. The appearance of Loakating Creek in Gordon's (1834:167) gazetteer shows that the local people also continued to employ spellings more closely resembling the name's originally documented form. The spelling of Lauhoekonatong Creek documented in 1725 (Moreau 1957:52) probably most closely approximates what Lockatong sounds like in the Munsee language.

LONETOWN (Fairfield County, Connecticut). The Lonetown Indian community was a home to Chickens and his people in the present-day town of Redding. Gradually selling their lands in the area, they finally exchanged their last hundred acres there for two hundred acres farther into the interior at Schaghticoke in 1748. A hamlet, a cemetery, and the Lonetown Marsh Sanctuary currently preserve the name in the town.

LOPATCONG (Warren County, New Jersey). Whritenour thinks Lopatcong sounds like a Munsee word, *lahpihtukwung*, "at the swift river." Lopatcong Creek is a nine-mile-long stream that rises on the west slope of Scotts Mountain in Harmony Township. From there it flows south into Lopatcong Township through Lows Hollow to the place where the Morris Canal joins it at Port Warren. Both waterways flow side by side into the town of Phillipsburg, only parting company when the creek turns to fall into the Delaware River two miles short of the canal's junction with the river across from the mouth of the Lehigh River. Present-day Lopatcong Township was named after the creek and was originally part of Phillipsburg Township. It became one of the towns formed out of parts of Phillipsburg not included when the latter place established itself as an independent municipality in 1863. Lopatcong Creek was first noted in a grant given to operate a ferry across a nine-mile-long stretch of the Delaware River between the mouths of what were referred to as the Lopetekong and Muskonetkong creeks that was first extended by a February 12, 1735, patent to New Jersey colonist David Martin (cited in a deed dated September 26, 1788, on file in the Sussex County Hall of Records, Deed Book A:271–74).

LUPPATATONG (Monmouth County, New Jersey). Luppatatong Creek is a four-mile-long mostly tidal stream that rises in the Mount Pleasant Hills in Holmdel Township. The creek flows north through Hazlet Township into wetlands at its junction with the Matawan Brook at Keyport. The place was first mentioned as Lupakitonge Creek in a colonial patent to land in the area dated June 20, 1687 (Whitehead et al. 1880–1931[21]:106). Luppatatong was a small shipbuilding community during the 1800s. As with Chingarora mentioned earlier, the reputation won by Luppatatong oysters grown in beds maintained in the waters around the creek's mouth brought the name to wider attention during the latter decades of the nineteenth century.

MACHACKEMECK (Orange County, New York). Whritenour thinks that this place name sounds much like a Munsee word for "red grounds." The name first appeared in Dutch records when local Indian leaders apologized for mischief caused by an "Indian from Mechagachkamic" near Manhattan at a treaty meeting held in New Amsterdam on July 19, 1649 (O'Callaghan and Fernow 1853–87[13]:25). Rectangular symbolic representations of longhouses indicating Indian town locations identified as Meochkonck and Mecharienkonck in the present-day Port Jervis area appear on Jansson-Visscher series maps produced after 1650; either, both, or neither may mark Machackemeck. On February 6, 1694, the name was more recognizably noted as Maggagamieck, a place located a half day's journey from Minnissinck (O'Callaghan and Fernow 1853–87[13]:98). Colonists used various spellings of the name to identify the Neversink River that joins the Delaware River at Port Jervis. Formerly given to village institutions such as the Dutch Reformed Church and the volunteer fire department, the name gradually fell out of use by the mid-1900s. The spelling of Machackemeck was selected recently to support a folk translation of "pumpkin ground" for the historic Magagkamack Churchyard. The cemetery contains the graves of more than three hundred area residents buried between 1737 and 1850 and represents the only current example of the place name on modern maps.

MACOPIN (Passaic County, New Jersey). Heckewelder thought the name of the place he identified as Makiapier Pond in New Jersey came from a Delaware word, *machkiabi,* "water of a reddish color." Cedars that may have given pond waters a reddish tint have long since been cut down. Although Whritenour finds Heckewelder's etymology credible, he thinks Macopin sounds much more like a Munsee word, *mahkwupiing*, "at the water of bears," perhaps in reference to the kind of place bears bathe, hunt, fish, or drink. Today, Macopin is the name of a river, a reservoir, and the hamlets of Macopin and Upper Macopin in West Milford Township. Macopin and Makpi Pond were first mentioned in survey papers dated 1753; it was subsequently noted as a "pond called Mekepien at a place called the Pleasant Ridge" on May 14, 1757 (New Jersey Archives, East Jersey Proprietary Survey Book S-4:110); and Makabien Pond is shown on a 1767 map of Bergen County. The name is spelled Macookpack on the 1799 Sauthier map. Today the pond whose name was spelled so many ways is called Echo Lake. The name of the Macopin River, which still flows into the lake, was never abandoned, however. The modern-day Macopin communities almost certainly adopted the name to commemorate a local sachem variously noted as Machopoickan and Mackpoekat in four Indian deeds to lands between the Rockaway River and the uppermost reaches of the North Branch of the Raritan negotiated between 1701 and 1702.

MACUNGIE (Lehigh County, Pennsylvania). Heckewelder thought Macungy sounded like a Delaware word, *machkúnschi*, "the harboring or feeding of bears." Whritenour thinks Macungie sounds more like a Northern Unami word, *machkawonge*, "red hill." Today, Macungie most prominently appears on maps as the name of two townships and a borough. The name first occurred as Macousie and Maquenusie in a road return dated March 27, 1735 (State of Pennsylvania 1851–53[3]:617). It next appeared as Mahquongee in the May 9, 1738, deed to one hundred acres of land within the bounds of the present-day borough of Macungie (Moravian Archives, Provincial Elders Conference Papers, Folio 6.4). Peter Miller renamed the place later established as the borough of Millersville after himself when he moved to the area in 1776. Millersville became a borough in 1857 and was only given back its current name of Macungie in 1875. The place originally named Macungie included land around the present-day borough set off as Macungy Township in 1752. The township was divided into Upper and Lower Macungie townships in 1832. The area in and around the townships subsequently became well known among archaeologists as a locale where Indians mined the distinctive glassy-brown jasper cherts that they chipped and flaked into sharp-edged stone tools.

MAHONING (Carbon County, Pennsylvania). Heckewelder thought that the Delaware words *mahóni*, "a deer lick," and *mahonink*, "at the lick," sounded much like the present-day place name. Whritenour thinks Heckewelder's etymology is correct. Today, the name Mahoning on Greater New York maps graces a creek, a township, a road, and the village of New Mahoning west of the Lehigh River at the eastern end of the valley between the Blue Mountain and Mauch Chunk ridges. Like most other Indian place names at the westernmost end of the Munsee homeland, Mahoning first came to colonial notice around the time of the 1737 Walking Purchase. It first appeared as Mohaining in a November 12, 1740, letter written for Indians protesting their eviction from lands they did not think were within the limits of the Walking Purchase (American Philosophical Society, Logan Papers [4]:71–72). The stream then appeared on the Evans map of 1749 in its present location as Mahoning Creek, a seventeen-mile-long brook that flows into the Lehigh River at Lehighton. Use of the name subsequently spread beyond the Lehigh Valley farther west into the more distant parts of Pennsylvania and adjacent sections of Ohio, where it adorns creeks and rivers and much else.

MAHOPAC (Putnam County, New York). Whritenour suggests that Mahopac sounds much like a Munsee word, *meexpeek*, "that which is a lot of water." Today, Mahopac is the name of a lake and several nearby communities, roads, and other places along the headwaters of the Muscoot River in the town of Carmel. The name appears to be a conflation of names from two sources. The earliest are Meconap and Mecopap, variant spellings of the name of an Indian mentioned in a deed to land in the area dated August 13, 1702 (Pelletrau 1886:Plate 2). The second source may be the name noted as Wakapa Creek on a map prepared for use in a local boundary dispute in 1753 (Library of Congress, Maps of North America 1750–89, 1083). Although the latter name almost surely referred to Wiccopee Creek, the map erroneously extended the route of the watercourse much farther south to land in the Hudson Highlands, very close to what on the map is an unnamed body of water that is most likely present-day Lake Mahopac. Whatever its original name, the lake was variously known as Great Pond and Hughson's Pond (for a settler who moved to its shores in 1740) throughout much of the colonial era. The name Mahopac began appearing with increasing frequency during and after the Revolutionary War, first as the name of the lake referred to by Spafford (1813:152) as Mahopack Pond, and later as the name of the falls at Red Mills southwest of the lake. It was finally noted in its present form as the name of the post office built in 1849 at the hamlet of Mahopac, which lay at the east end of the lake. Today, use of the name has expanded to include such nearby communities as Mahopac Falls, Mahopac Mine, Mahopac Point, and West Mahopac.

MAHORAS (Monmouth County, New Jersey). Whritenour thinks it is very likely that Mahoras is an anglicized version of a Northern Unami word, *mahales,* "flint or chert." Today, Mahoras functions as the name of a creek and as a street name. The creek rises in Holmdel Township just east of the Garden State Arts Center. From there it flows into Middletown Township north to its junction with Waackack Creek at Philips Mills. The name first appeared in a series of three Indian deeds to lands in the area signed between 1676 and 1677. The first, dated February 3, 1676, mentioned a brook called Hepkoyack or Mohoras (New Jersey Archives, Liber 1:257[82]–256[83]). The second, dated September 29, 1676, included a run or swamp called Mohorhes (Monmouth County Records, Deed Book B:33–35). The last, dated August 10, 1677, mentioned a place called Mohoreas (New Jersey Archives, Liber 1:105).

MAHWAH (Bergen County, New Jersey, and Rockland County, New York). Whritenour affirms that earlier translations suggesting that Mahwah was a Munsee word for "assembly" (he proposes **maaweewii*) were correct. Today, Mahwah is the name of a township and a village in New Jersey. It is also the name of the nine-mile-long stream that flows below the southeastern slope of the Ramapo Mountains from its headwaters in the town of Haverstraw, New York, into the town of Ramapo and across the state line to its junction with the Ramapo River at the village of Mahwah. The gap the Indians called Opingua as early as 1653 has since gone by the names Ramapo Clove and Ramapo Pass. The earliest known reference to an "Indian field called Maweway" occurred in the first Indian deed to land in the area signed on November 18, 1709 (New Jersey Archives, Liber I:319–21). Copies of the survey return for the deed completed six months later used variations on the spellings Mawaywaye and Maygahtgayako to identify the Indian field at the flats by the mouth of the present-day Mahwah River (ibid.:317–19, 321–22). A nearby tract just to the north in New York sold on April 23, 1724, identified Mawewieer as a southern boundary marker for the property (Budke 1975a:111A). Palatine German refugees began settling in the area a decade earlier after first landing in New York in 1710. Both they and Dutch settlers already living there established the foundations of the agricultural economy that sustained farmers living in and around local hamlets initially called "the Island," Hopper's, and Baldwin Mills (Bischoff and Kahn 1979:48–49). Residents at these and nearby communities selected the name Hohokus when they established a township of their own in the area in 1849. Mahwah had long been used informally as a local name at its present locale when Erie Railroad managers selected it as the name of the station opened in the hamlet in 1909 (Wardell 2009:58). Industries established at West Mahwah after 1900 expanded rapidly. The name Mahwah was ultimately chosen to replace Hohokus for the township in 1944.

MAMAKATING (Sullivan County, New York). Whritenour thinks Mamakating sounds similar to a Munsee word, *mahmaxkatun*, "red mountain." Today, Mamakating is the name of a town located at the southeastern end of Sullivan County. Colonists purchased much of the land in the present-day town in an Indian deed finalized on June 8, 1696 (Ulster County Records, Deed Book CC:145). The name Mamakating itself initially appeared in colonial documents recording Indian challenges to the bounds of the 1696 deed, first as Mammekotton in the complaint made on August 22, 1722, and then as Mamecatten and Memekitton in the April 21, 1730, judgment settling the dispute (both in Special Collections, Alexander Library, Rutgers University, Philhower Collection Transcripts). Early European settlement in the area centered around three places where fortified houses were erected in 1753. Fort Westbrook was built on the present-day Sullivan-Orange County line. Fort Devans was located above modern-day Wurtsboro near the banks of what was then called the Mamakating River (today's Basher Kill). Fort Roosa was built at Roosa's Gap, through Shawangunk Mountain ridge north of Wurtsboro. Population gradually began to concentrate along the lowlands just south of Fort Devans at a place originally called Mamakating Hollow during and after the Revolutionary War. The local population grew large enough to warrant incorporation of a town that residents named Mamakating in 1788. People living in the western part of this new town split off to form one of their own called Thompson in 1803. The growth rate of the village Mamakating (then also called Rome) was sufficient to warrant construction of a church in 1805 and a post office in 1817 (Kaiser 1965). Construction of the Delaware and Hudson Canal changed everything in the village, including its name, which was formally rechristened Wurtsboro in 1828 to honor the Canal company's founder who had briefly lived in the community. Today, Mamakating is preserved as the name of the town, as Mamakating Park, a residential community originally planned to be a resort destination just west of Wurtsboro, and as the Mamakating Town Park in the village of Bloomingburg.

MAMANASCO (Fairfield County, Connecticut). Whritenour suggests that Mamanasco may come from a Munsee word, *(eenda) *mahmunaskwahk*, "[where] grass is repeatedly gathered." Today, Mamanasco is the name of a lake, a road, and a residential subdivision in the town of Ridgefield. It first appeared in colonial records as "a place called Mamanasquag" in a September 30, 1708, deed to land in the area and was mentioned as Mamanasco Hill in a later conveyance dated November 22, 1721 (Teller 1878:3–6, 23–24). The small pond that today bears the name was created to provide the water to turn the wheel of the Mamanasco Mill, which operated at the locale during the 1700s and 1800s. Construction of the Mamanasco Lake Park development began on the banks of the lake in 1957.

MAMARONECK (Westchester County, New York). Nora Thompson Dean thought Mamaroneck sounded similar to the Unami word *mehëmalunèk*, "place to dance." Whritenour notes that the name sounds much more like a Munsee word, **maamaalahneek*, "striped stream." Today, Mamaroneck is the name of a river, a reservoir, a harbor, a neck, a town, a village, and much else in Westchester. The name appeared with corresponding frequency during colonial times following its initial mention as the Mamaronock river, neck, and tract in the first Indian deeds selling land in the area signed on June 6 and September 23, 1661 (in Robert Bolton 1881[1]:466; Palstits 1910[2]:648–54). Indian sales of lands along the upper reaches of the river helped Caleb Heathcote assemble his Scarsdale Manor by the end of the first decade of the 1700s. One of these deeds contained a clause "reserving liberty for [the Indian signatory's] use, such whitewood trees as shall be found suitable to make canoes of" (in Robert Bolton 1881[1]:362–63). Mamaroneck and Larchmont were combined to form the town of Mamaroneck in 1788. The hamlet of Mamaroneck's location astride both banks of the Mamaroneck River that became the town's border with the neighboring town of Rye ultimately resulted in the creation of a village under the jurisdiction of two towns when the hamlet was incorporated as Mamaroneck in 1895. The Mamaroneck River that continues to flow through the village remains a small five-mile-long stream paralleling Mamaroneck Avenue whose waters rise just one mile south of the Old Mamaroneck Road neighborhood in White Plains. From there it flows southeast past Mamaroneck Neck into Long Island Sound at Mamaroneck Harbor. A fork of the Old White Plains Road still splits off into Mamaroneck Avenue at the Mamaroneck village center, only to turn into the Mamaroneck Road that runs toward Scarsdale once it crosses the Mamaroneck town boundary.

MANALAPAN (Middlesex and Monmouth counties, New Jersey). Whritenour notes in Boyd (2005:441) that it is almost certain that Manalapan is a Northern Unami place name meaning "wild onion patch," from *men*, "collected together in a place," and *ulepin*, "wild onion." Today, Manalapan is the name of a brook, a township, a village, and a local street. Manalapan Brook is a sixteen-mile-long stream that rises in Millstone Township and flows north through Manalapan Township into Middlesex County. In Middlesex the stream passes into Monroe Township through Jamesburg and Outcault (once known as Weequehela's Upper Saw Mill) to Spotswood. From there it becomes the South River (formerly known as the Manalapan River), a tidal stream that flows through Old Bridge into Raritan Bay between Matawan and Keyport. Manalapan has been on colonial maps since a place called Manoppeck was mentioned in the August 15, 1650, Indian deed to land by Raritan Bay (New Jersey Archives, Liber I:6–7). The name first came into frequent use in its more recognizable modern form in a series of Indian deeds signed in the area between 1696 and 1741. Weequehela (Heckewelder thought his name came from a Delaware word meaning "to be fatigued"), a sachem known to colonists as the Indian king of New Jersey who signed many of these deeds, lived along the brook with his people. They remained there until all but a few relocated to the Forks of the Delaware in 1727, soon after Weequehela was tried and hanged for killing a local settler while drunk (Grumet 1991; Wilk 1993). Most returned to the Manalapan Valley after being evicted from the Forks following the Walking Purchase of 1737. Many of these people were converted to Christianity by the Presbyterian missionary brothers David and John Brainerd. Most finally left the area and moved to the Brotherton Reservation established at Edgepillock in 1759.

The small hamlet of Manalapan, later called Tennent, just southeast of Englishtown, grew slowly in the decades following the final Indian abandonment of the area. The locale became a Revolutionary War battleground when American and British troops faced off at the Battle of Monmouth on June 28, 1778. A post office given the name Manalapan had already been opened at what was then called Tennent when area residents named their newly erected township Manalapan in 1848. The present-day community of Tennent Station on the line of the now-defunct Freehold and Jamesburg Agricultural Railroad built across the township after 1851 was originally named Manalapan Station.

MANASQUAN (Monmouth County, New Jersey). Nora Thompson Dean thought that Manasquan sounded much like a Southern Unami word, *mënàskung,* "place to gather grass." Whritenour suggests that more likely it may mean "stream or place of the second crop" from the Northern Unami words *manaskw,* "second crop," *-an,* "stream," and *-ung,* "place of." Today, Manasquan is the name of a river, a reservoir, an inlet, a borough, and several other places. The Manasquan River is a twenty-seven-mile-long stream that flows through several townships, rising first in Freehold, flowing through Howell, and descending into Wall Township. There it widens into the tidal Manasquan Inlet estuary before flowing out to sea through a canal cut across the barrier beach separating the mainland from the Atlantic Ocean. The name first appeared in colonial records in a license issued on July 9, 1685, granting settlers permission to purchase land at Manisquan from the Indians. They signed a deed to land "commonly called Squancum . . . [on] the Mannusquan River" four years later

(Monmouth County Records, Deed Book B:92–93). Settlement subsequently centered around what local residents called Squan Village until they changed its name to Manasquan when they formally incorporated their community as a borough in 1887. The name also adorns several roads and neighborhoods around the borough, as well as the state of New Jersey's 744-acre Manasquan River Wildlife Management Area at the head of the inlet and the Manasquan Reservoir built in 1990. Located some distance from the river, the reservoir is connected to the river by an aqueduct. Pumps fill and drain the reservoir with river water as needed. The Monmouth County Park Commission manages the 1,204-acre Manasquan Reservoir County Park located on land around the impoundment and pump house.

MANETTO (Ocean County, New Jersey; Nassau, New York, and Putnam counties, New York). Whritenour thinks the names of places named Manetto, Manitou, and Minetta can trace their origins to a Munsee word, *méeneet*, "drunkard," whose surface sound similarity to the otherwise unrelated Manitou, from the Munsee word *manutoow*, "spirit being," appears to be the common thread that links variants of these names found at several places in the present-day Greater New York area.

In today's Nassau County, for example, residents called their community Manetto Hill after the eminence first identified as Mannatts Hill in a local Indian land sale made on October 18, 1695 (Cox 1916–40[4]:513–14). They continued to use the name until 1885, when postal authorities rejected a request that they give it to the village's newly built post office on the grounds that it was too similar to another name elsewhere in the state. Residents responded by giving the name Plainview to the new branch and the community that it served. Today, Manetto Hill survives as a road name in Plainview and, in slightly different form, as Greenwich Village's Minetta Lane and Place. Both roadways mark the former course of Bestavaer's Killjte, a now-buried stream renamed Minetta Brook at the suggestion of local historian Reginald Pelham Bolton (1905:163) around the turn of the twentieth century. Evidently unrelated Lake Manetta, a millpond built in 1883 to provide water needed to power the Bergen Iron Works in Lakewood, New Jersey, is probably an Italian or other European surname.

MANHATTAN (New York County, New York). Manhattan is, by any measure, one of the world's most instantly recognizable place names. The legend of the twenty-four-dollar price paid for the island has played a major role in securing the name's international notoriety. It was the first Munsee word recorded by colonists, written as Manahata in 1610 in one of Hudson's sailor's journals, and, on a map discovered in the early twentieth century in Spanish archives, as Manahata on the west side of the present-day Hudson River and Manahatin on its east bank (no island is depicted). This map, represented and regarded since its discovery as a spy map based on information culled from Henry Hudson's since-lost journal, may be a forgery (see Allen 2006; but also see Goddard 2010:279n.6). Manhattan is also one of the very few Munsee place names that has never been removed from regional maps. Even New York, whose history is inseparably intertwined with Manhattan's, only first appeared on regional maps in 1664 and, unlike Manhattan, was removed by Dutch authorities who briefly reconquered the region in 1673 (their replacement, New Orange, only lasted a year). The histories of Manhattan and New York are so closely connected that a history of one is much like that of the other.

No records directly document the 1626 Indian sale of Manhattan. The purchase and its sixty-guilder purchase price in trade goods were both mentioned briefly in a letter written in 1626. A record of some sort was evidently shown to Indians claiming land in Harlem in 1670 which showed that the place had been "bought and paid for forty-four years ago" (Palstits 1910[1]:48). The trail ends there for those interested in the history of Manhattan as an Indian place name. All else is etymology. Translations have ranged from William Wallace Tooker's (1901) "hilly island" to Heckewelder's *Manahachtanienk*, "island where we all became intoxicated." Heckewelder, who tended to associate Delaware words beginning with *man* or *mana* with drinking, based the latter translation on a historically undocumented Delaware tradition recounting their ancestors' reaction after seeing one of their number fall dead and revive after drinking from a cup proffered by a Manitou said in the story to be Henry Hudson. Two other translations of Manhattan also come from Delaware sources. Personally, I have always favored the simple translation "island," based on the Unami word *Mënating*, suggested by Nora Thompson Dean and others. Linguist Ives Goddard (2010) champions nineteenth-century Canadian Munsee Delaware writer, scholar, and fluent Munsee speaker Albert Anthony's etymology of *Man-ă-hă-tonh*, "the place where timber is procured for bows and arrows" that Anthony stated was obtained from traditionalists. Goddard (2010:288) uses Heckewelder's (1841:73) Mahican attestation of *Manhatouh* as a name "owing or given in consequence of a kind of wood which grew there, and of which the Indians used to make their bows and arrows" to independently confirm Anthony's etymology. Whatever its original form or meaning, documented examples show that Delaware-speaking people have given three interpretations of the name at one time or another to people expressing interest in its origins over the course of the past few centuries.

MANITOU (Putnam County, New York). Whritenour notes that modern spellings of the place name Manitou often represent associations with the Munsee word *manutoow*, "spirit being." The name occurs widely in many forms throughout North America. Places called Manido Falls, Manito, and Manitou in Michigan; Manitou Springs in Colorado; Manitowac and Minito in Wisconsin; and Moniteau, Missouri, all trace their etymological origins to words for spirit in other Algonquian languages.

A cluster of places named Manitou and Manitoga in the village of Garrison in Putnam County, New York, bear the name of an Indian man mentioned in a deed to land in the area dated July 13, 1683, as Mantion (Budke 1975a:53–55) and as Manito on another deed, signed in 1710, to land some thirty miles farther west at Ramapo in today's Rockland County (New Jersey Archives, Liber 1:317–19). Local residents, evidently believing that the man's name meant "spirit," gave it to their village, to a nearby 774-foot-high mountain, and to a point of land jutting into the Hudson River in the town of Phillipsburg. Pioneering industrial designer Russel Wright gave an altered form of the name of his own invention, combining Manitou with the last part of the name of fashionable Saratoga, to his seventy-five-acre Manitoga estate built in the village of Garrison in 1941. Manitoga presently houses the Russel Wright Design Center, next door to the 137-acre Manitou Point Nature Preserve.

MANNAYUNK (Orange County, New York). Five-mile-long Mannayunk Kill flows into the west bank of the Walkill River just above the present-day village of Montgomery. Surveyor Peter E. Gumaer noted the stream as Mononcks Kill on May 4, 1809, and as Menonks Kill six days later in his survey diary (Ogden 1983:43, 45). An Ulster County trader listed a leader identified as Manonck and his kin among his most active clients between 1717 and 1725 (e.g., Waterman and J. M. Smith 2011:5, 6, 62). References to Manonck's burial and final account balances in 1726, penned around the same time that the sachem Nowenock last appeared in colonial documents, suggest that the similarly spelled names may have belonged to the same person. On January 21, 1777, early Menissink settler Jacob Westfall remembered Menonck as one of several Indians he knew in his youth at and around Kishiston (Cochecton) (New York State Library, Indorsed Land Papers [28]:71). Local residents evidently gave a slightly respelled version of what was to them the more familiar place name Manayunk (Pennsylvania's Schuylkill River and the present name of a Philadelphia suburb that Heckewelder thought meant "drinking place" in Delaware) to the stream as memories of Manonck faded during the nineteenth century.

MANSAKENNING (Dutchess County, New York). This place name was first mentioned as "a fresh meadow called Mansakin" in the 1686 Beekman Patent to land in present-day Dutchess County. Writing around the turn of the twentieth century, Beauchamp (1907:55) noted that the place formerly called Mansakin was then known as Jackomyntie's Fly. A few years earlier, the builder of a Dutchess County estate who resurrected the name from local records decorously altered it to Mansakenning before giving it to his new farm just south of the village of Rhinebeck around 1903. The well-preserved Mansakenning Farm estate, listed in the National Register of Historic Places in 1987, retains the name to the present day.

MANUNKA CHUNK (Warren County, New Jersey). Nora Thompson Dean thought that Manunka Chunk closely resembled a Southern Unami word, *mënàngahchung*, "where the hills are clustered." She also thought that its earliest orthography, Penungachung, sounded like another southern Unami word, *pënaonkòhchunk*, "place where the land slopes downhill." Whritenour suggests a Munsee phrase, *peenang wahchung*, "one who looks at the hills." Today, Manunka Chunk is the name of a mountain, an island, a road, and an unincorporated community in White Township. John Reading, Jr. (1915:40–42), used a variety of spellings for the place he recorded as the Penungachung Hills in journal entries made between May 18 and 21, 1715. Anthony Dennis noted islands and a mountain he identified as Monongachunk on his 1769 map. A Delaware, Lackawanna, and Western Railroad man probably concocted the spelling Manunka Chunk about the time the company was laying track and cutting a 975-foot-long tunnel through the mountain for its main line between 1854 and 1856. Transcription of Reading's Penungachung into Manunka Chunk almost certainly was intended to create a link in people's minds connecting the locale with the very popular Mauch Chunk resort area in the Lehigh Valley. Travelers from the lower Delaware Valley traveling north on the Pennsylvania Railroad's Belvidere-Delaware Division on their way to Mauch Chunk and other resorts stopped off at Manunka Chunk to switch trains following completion of the junction platform at the locale in 1876. More than a few returned for a more extended visit after taking in the area's scenery while they waited. One of these visitors was a British travel writer who, in a book recounting his American journey, noted the connection between Reading's Penungachung and the Manunka Chunk railroad name (G. Wright 1887:170–71). Manunka Chunk's subsequent emergence as a Delaware Water Gap resort destination convinced railroad executives to order the building of a full-service passenger station at the locale in 1899. Many of the tourists from New York and Philadelphia stepping off the train at the new station went straight to the ferry that carried them across the Delaware River to the Manunka Chunk House resort hotel on nearby Manunka Chunk Island. The flood of 1913 carried off both the station and the ferry slip. Utterly destroyed, neither was rebuilt and the cut-off Manunka Chunk House was soon abandoned. Vandals finally burned the derelict hotel in 1938. The railroad ultimately followed the local tourist hostelry industry into oblivion in the decades preceding the line's abandonment in 1970. Today, little more than some road signs and a few businesses mark the place present-day maps still identify as Manunka Chunk.

MANURSING (Westchester County, New York). Whritenour finds that Manursing sounds much like a Munsee word, *munusung, "at the island." It was first identified as Manussing, an island just across from Peningo on the mainland in the present-day city of Rye sold in a June 29, 1660, Indian deed (in Robert Bolton 1881[2]:130). It remained on colonial maps and was retained by a succession of buyers through the nineteenth and into the twentieth centuries. Landfill, dumped to provide secure foundations for Rye Playland, built on Manursing Island in 1928 and designated a National Historic Landmark in 1987, has all but obliterated traces of the original shoreline. Today, the name Manursing remains on maps and street signs on and around the former island within the city limits of Rye, a free-standing municipality that separated from the town of Rye in 1942.

MASHIPACONG (Sussex County, New Jersey). Whritenour thinks Mashipacong sounds like the Munsee words *machiipeekwung, "place of bad water," and *mahchupahkwung*, "place of white pine trees." Today, Mashipacong is the name of an island, a pond, and the Old and New Mashipacong roads that connect both places in Montague Township. The name made its initial appearances in colonial records in entries spaced a week apart in John Reading, Jr.'s survey book. On June 24, 1719, Reading (1915:95) noted "a large piece of lowland called Machippacong." One week later on July 1 (ibid.:99), he discovered that it was actually a large island on the Jersey side of the river. Mashipacong next appeared when local mill owner John Rutherford gave the name to the head pond he created by damming streams flowing into a bog located on a stretch of level land atop the Kittatiny Ridge to provide water to power two of his nearby mills in 1848. Local residents subsequently used the name when talking about the road running between the island and the pond. Mashipacong Island remains a frequently flooded untenanted place mostly used as a farm field. Mashipacong Pond subsequently was purchased by a group of local sportsmen who opened their private Mashipacong Club in 1901. The Civilian Conservation Corps gave the name to a still-standing stone shelter it built at a dramatic lookout point along the Appalachian Trail just east of the pond in 1936. Two years later, philanthropic tobacco fortune heiress Doris Duke bought the land around the pond and leased it to *Life* magazine for the site of one its Fresh Air camps operated for needy New York schoolchildren. Duke finally donated the property to the Nature Conservancy in 1991 on condition that it reserves two hundred acres for the multicultural Trail Blazer's Camp at the site of the former Fresh Air facility. Today, the Conservancy preserves the ecosystem supporting the unique northern boreal bog within its thousand-acre Mashipacong Bog Preserve.

MASONICUS (Bergen County, New Jersey). Whritenour thinks Masonicus sounds like a Munsee cognate of a Northern Unami word, **mallsannukus* (the L is not voiced and might be missed by a colonial translator), "little flint or chert arrowhead." Today, Masonicus is a street and neighborhood name in the Bergen County borough of Mahwah and the name of a five-mile-long brook that runs through Mahwah from the Masonicus neighborhood to its junction with the Mahwah River at West Mahwah. The name first appeared late in the colonial era as a place variously called Massanuckes and Messankes in local road returns made between 1754 and 1769. It has been on maps of the area ever since.

MASPETH (Queens County, New York). This Munsee word is sometimes translated, often ironically, as "bad water." Today, Maspeth is the name of a neighborhood and a long-polluted stream at the head of Newtown Creek in the borough of Queens in New York City. Dutch West India Company officials purchased land bordered on the north by what was called the thicket of Mespaetches when they bought all of present-day Williamsburg from its Indian owners on August 1, 1638 (Gehring 1980:8–9). Since colonial times the name of Maspeth has continuously been on local maps in one form or another as the name of the small community at the head of Maspeth Creek (Riker 1852). Use of the name Maspeth for the whole of the modern-day Newtown Creek had shrunk to the present-day creek and the "west branch of Marshpath kills called Quandus Quaricus" (Heckwelder gave a dictionary translation of *gunig quingus* as a Delaware word for "grey ducks"; the stream is today known as English Kills) by July 9, 1666, when sachems confirmed their April 12, 1656, sale of the tract that Dutch colonists first called Middelberg, between the Maspeth and Flushing creeks. English settlers later gave the name Newtown to the place and its nearby creek (Palstits 1910[1]:235–37).

MASSAPEQUA (Nassau County, New York). Nora Thompson Dean thought Massapequa sounded much like a Southern Unami word, *mësipèk*, "water from here and there." Today, Massapequa is the name of a creek, a lake, and a cluster of municipalities at the southeastern corner of the town of Oyster Bay. Massapequa has been on maps since the sachem of Marossepinck signed the January 15, 1639, treaty deed that granted the Dutch West India Company the sole right to purchase Indian land in western Long Island (Gehring 1980:9). Colonists initially considered Massapequa a remote backwater too close to the Indian community centered at Fort Neck. This did not stop them from purchasing the Massapeaque Meadows as pasture land from Tackapausha on March 17, 1658 (Cox 1916–40[1]:347–49). Montaukett sachem Wyandanch claimed an interest in this land sale, one of several he involved himself in between 1657 and 1658 along the present-day Nassau-Suffolk county line. Despite Wyandanch's death a year later, claims made by his descendants (and of hopeful eastern Long Island colonists claiming his lands) continued to roil local waters for many years (Strong 1997:221–30). Colonists finally started pressing into the area during the early 1690s. Negotiating several deeds, they finally acquired the last Indian rights to lands at Fort Neck in 1697. Although Indian people continued to live at the locale at one time or another during succeeding years, they located most of their remaining settlement centers in places in and around two small reservations at the head of Hempstead Harbor set up for their use by provincial authorities in 1687.

Massapequa remained a small farming village until the Long Island Railroad opened its South Oyster Bay Station in the hamlet in 1891. Developers soon began selling house lots, many of which were purchased by well-to-do second-generation Irish immigrants. Massapequa grew rapidly during the following century. One particularly fast-growing subdivision called Massapequa Park determined to control its own development incorporated itself as a separate village in 1902. Other children of immigrants and their descendants still move to the increasingly densely populated cluster of communities in and around present-day Massapequa.

Today, Massapequa Lake serves local communities as a major recreational focal point. The story of the lake began when the city of Brooklyn enlarged the old mill dam on Massapequa Creek to create the Massapequa Reservoir in 1890. A conduit connected the reservoir to the Brooklyn Waterworks pumping station four miles farther east in Freeport. From there water was pumped through a larger water tunnel past present-day Aqueduct Racetrack (hence the name) to the Ridgewood Reservoir atop the terminal moraine on the Brooklyn-Queens border. Little used after Brooklyn gained access to the Croton System following its incorporation into the city of New York in 1898, the Massapequa Reservoir continued to serve as a backup supply source until the 1970s. Today, the 423-acre Massapequa Preserve taking in the old reservoir and a four-mile-long stretch of Massapequa Creek is managed by the Nassau County Department of Parks, Recreation, and Museums.

MASTHOPE (Pike and Wayne counties, Pennsylvania). Heckewelder thought the name Masthope sounded similar to Delaware words that he spelled *maschápe* and *mashapi* and that he translated as "glass beads." Today, Masthope is the name of a creek and a village located where the creek flows into the Delaware River in the town of Lackawaxen. Six-mile-long Masthope Creek rises just east of the hamlet of Beach Lake in the Wayne County town of Berlin. Anthony Dennis located Masconos Creek above the legend "and once his settlement" at the locale on his 1769 map. Initial settlements made by New Englanders in the area around this time failed. Local tradition holds that loggers guiding rafts down the Delaware used the stretch of quiet water below the small fan of gravel outwash at the mouth of the creek as a rest stop during the final decades of the eighteenth century. Tradition holds that people moving to the small community that grew up at the creek mouth adopted its name partly in hopes that sales of masts made from local logs would bring prosperity. Development, however, began in earnest only after the New York and Erie Railroad laid tracks on the bridge built across the mouth of Masthope Creek in 1848. The post office that opened at the optimistically named Delaware Bridge community across the river in 1849 transferred operations to Masthope in 1855 when the hoped-for span did not materialize. The village and its nearby creek both continue to be known as Masthope to the present day. The name originally belonged to a local Indian whom the settlers knew as Mastewap. Seventy-one-year-old settler Johannis Decker affirmed in a deposition taken in 1785 (Snell 1881:368) that he had known Mastewap and other Indians "at Coshecton, Shohacan, and Cookhouse [present-day Deposit, New York]" when they signed the deed that New Englanders used to claim land above the Delaware Water Gap on May 6, 1755 (Boyd and Taylor 1930–71[1]:260–71). Mastewap's name was variously spelled Amossowap, Mastohop, and Malonap in other records documenting land dealings in the area between 1746 and 1761.

MATAWAN (Monmouth County, New Jersey). Rementer (in Boyd 2005:425) suggested that Matawan sounded much like *matawonge*, a Southern Unami word for "bad river bank," consisting of *mat*, "bad," and *awonge*, "stream bank." Whritenour, in the same source, added that *awonge* could also mean "hill." Whritenour more recently thinks that the name also sounds very much like a Northern Unami word, **matawan*, "bad fog." Today, Matawan is the name of a brook, a lake, a dam, a point, and a borough. The name first appeared as Mattawane Creek in a May 12, 1688, confirmation of a land purchase made by

Scottish Quaker and Presbyterian settlers in the area (Whitehead et al. 1880–1931[21]:119). It was next mentioned in an October 10, 1690, Indian deed to land between what were identified as Matewan and Mohingsunge creeks (Monmouth County Records, Deed Book C:163). In 1701 proprietary authorities formally granted village status to the community they named New Aberdeen after the town of Aberdeen in Scotland. Residents began calling the place Mount Pleasant as new immigrants from other places later moved to the area. They also changed the name of the community boat landing place originally called Matawan Point to Middletown Point by the early 1800s. Local residents incorporated themselves into a new township they named Matavan in 1857. Eight years later, the Middletown Point post office at the north end of the township changed its name to Matawan, an older spelling of the name. By 1885 confusion caused by the duplication of the names in the same place galvanized residents into action. They agreed to change the spelling of the township name back to Matavan and renamed the old Middletown Point village Matawan when they declared their community a freestanding borough in 1885. Sometime thereafter, the neck known as Middletown Point also regained its old name of Matawan Point. A little less than a century passed before naming arguments again broke out. Township authorities determined to increase the visibility of the locality's lagging name brand resurrected the old name Aberdeen in 1977. Local sources indicate that township residents hoped that this change, which advanced the township's name to a first-place position in listings of New Jersey municipalities, would increase its visibility with state legislators and funding agencies. Today, Matawan continues to be the name of the five-mile-long Matawan Brook, Matawan Point, the borough of Matawan, and Matawan Lake, formed by damming the Matawan Creek's Gravelly Brook tributary just below the borough line in 1923.

MATCHAPONIX (Middlesex and Monmouth counties, New Jersey). Matchaponix Brook rises at the confluence of the Topanemus and Weamaconk brooks just east of the village of Englishtown in Monmouth County's Manalapan Township. From there it flows north into Middlesex County where it becomes the border between Old Bridge and Monroe townships. It then joins with the Manalapan Brook at the borough of Spotswood (formerly known as Weequehela's Lower Saw Mill) to form the South River. Matchaponix first appeared in an Indian deed dated October 10, 1690, as Mechaponecks, the home of a man identified as Ilhosecote (Irooseeke, namesake of Iresick Brook) and two other sachems selling land between the Matawan and Mohingson creeks (Monmouth County Records, Deed Book C:163). The name was mentioned in deeds to two tracts subsequently signed by Weequehela, noted elsewhere as a son of Irooseeke, the first a deed of sale to land along what was called the Machiponix River on March 10, 1702 (New Jersey Archives, Liber H:220–21), and the second a confirmation of the 1690 purchase signed on August 1, 1716 (Monmouth County Records, Deed Book E:197). Matchaponix Brook has been continuously mentioned in local records by these and other spellings since colonial times. The tiny community that rose up along a several-mile-long stretch of land on the west side of the brook during the mid-1700s took its name from the stream. The village became part of Monroe when the township was incorporated in 1838. It was also around this time that the northern part of the Matchaponix community changed its name to Texas to honor the newly independent Lone Star republic; the lower portion of the hamlet retains its original name to the present day.

MATINECOCK (Nassau County, New York). Today, Matinecock Point and the village of the same name are located at the north end of the town of Oyster Bay. The name first appeared in April 1644 when the sachem of a place identified as Matinnekonck sued Dutch authorities for peace on behalf of his community and those of nearby Marospinc (Massapequa) and Siketauhacky (Setauket) after colonial troops destroyed two of their towns (O'Callaghan and Fernow 1853–87[14]:56). Both Indians and colonists subsequently referred to the broad expanse of Long Island Sound coastline on the island's north shore between Hempstead Bay and Cold Spring Harbor as Matinecock Neck. Since then, Matinecock has been attached to nearly every conceivable place in the area capable of bearing a name at one time or another. At present, Matinecock most conspicuously appears on maps as the name of the small village that was incorporated in 1928 just south of Locust Valley where the Society of Friends built their Matinecock Meeting House two centuries earlier. The name also adorns Matinecock Point, a neck of land that juts out into Long Island Sound a few miles north of the village.

MATTANO (Union County, New Jersey). Today, Mattano Park is a recreational facility managed by the Union County Department of Parks and Community Renewal in the city of Elizabeth. The Union County Park Commission began development of Mattano Park just below the Elizabeth River Parkway during the late 1930s. They named the park in honor of Mattano, a close relative of Ockanickon, Irooseeke, and several other leaders of mixed Munsee- and Northern Unami–speaking communities located across central New Jersey. Mattano became an influential sachem in his own right in predominately Munsee towns in Brooklyn and Staten Island and around Raritan Bay. Mattano signed the 1664 Elizabethtown Purchase and participated in several other land sales around New York Harbor made between 1649 and 1665.

MATTAWANG (Somerset County, New Jersey). Whritenour suggests that Mattawang sounds very much like a Northern Unami word, **mateewamung*, "at the bad river flats." Today, Mattawang is a street name and the name of a golf club and course. The name first appeared in two Indian deeds to land in Monmouth County signed in 1665, first as Mattawomung on April 7 and then as Mattawamung on June 5 (Municipal Archives of the City of New York, Gravesend Town Records:73–74). It was more specifically identified as "Matawong alias Milstone River" in the August 22, 1681, Indian deed to land along its course in Somerset County (New Jersey Archives, Liber 4:2). The Mattawang Golf Club in Belle Mead Township adopted the name in 1960 after a club member found it in a book in the local library. It subsequently became the new name of the former Pike Brook Country Club when the Mattawang group acquired the property in 1993. It has more recently been adopted as the street name Mattawang Drive in a subdivision located a couple of miles east of the Millstone River at the junction of South Middlebush and Suydam roads in Franklin Township.

MATTEAWAN (Dutchess County, New York). Matteawan is currently the name of a road and a state hospital in the city of Beacon. The name first appeared in the August 8, 1683, Indian deed to land between the present-day Fishkill and Wappinger creeks as a boundary point on "the south side of a creek called the Fresh Kill, and by the Indians Matteawan" (Hasbrouck 1909:35–37). Local residents had been casually applying the name to Fishkill Creek and the nearby Hudson Highland ridge (today's Beacon Mountains) when the directors of the newly formed Matteawan Manufacturing Company acquired land at Clay Mills at the falls of the Fishkill about a mile from the Hudson River in 1814. They soon named their new company town Matteawan. The locale grew large enough to support a post office of its own by 1849 (Kaiser 1965). The institution today known as the Matteawan State Hospital for the Criminally Insane (sometimes spelled Matawan) was built near the village in 1892. Use of Matteawan as a village name began to diminish after its residents joined with those in neighboring communities to form the city of Beacon in 1913. Today it survives as the name of the state hospital and the road that passes through the former location of the old Matteawan community on its way to downtown Beacon. Matteawan is also the source of names of places in other states, whose various spellings include Matteawan, Matewan, and Mattawana.

MEAHAGH (Westchester County, New York). The Knickerbocker Ice Company built and named Lake Meahagh in the town of Cortlandt in 1855. Based in Rockland County, the company became the largest supplier of ice to the New York City market when ice blocks provided the sole coolant for preserving food in ice chests and boxes. During its heyday in the 1880s, the company owned several lakes and used 1,000 horses to draw 500 wagons that annually hauled 40,000 tons of ice to the city. Adoption of electrified refrigeration gradually put the company out of business by 1924. Company officials selected the local Indian place name Meahagh as it was spelled in Robert Bolton's (1881[1]:86–87) published transcript of Stephanus van Cortlandt's August 24, 1683, deed to Indian land in the area. Examination of a manuscript copy of the deed on file in the New York State Library (Deed Book G:26–27) shows that the name originally appeared as Meanagh, a word that Whritenour thinks sounds much like a Munsee word, *meenaxk*, "fence or fort." Recently the town of Cortlandt retained Bolton's spelling when it established Meahagh Park on land surrounding the lake.

MERRICK (Nassau and Queens counties, New York). Whritenour thinks Merrick sounds like a Munsee word, **muluk*, "snow goose." Today, Merrick is a municipality, road, park, and waterway name adorning places located along the western reaches of Long Island's south shore. The name first appeared when sachems from "Mesapeage, Merriack, or Rockaway" sold half of the Great Meadows grasslands at Hempstead to colonists on November 13, 1643 (O'Callaghan and Fernow 1857–83[14]:53). Sachems representing the three communities appeared in other records selling or contesting sales of lands in what is now the southern half of Nassau County throughout the remainder of the seventeenth century. The complex maze of Merrick's grassy inlets provided secure campsites for Indian travelers and concealed lairs of privateers sailing out into the Atlantic through the Merrick Gut up to the end of the War of 1812. Development in the area began only after the Long Island Railroad ran its mainline tracks across the center of the island during the mid-1860s. Methodists coming by wagon and train established their Long Island Camp Meeting Association grounds at what soon became known as North Merrick in 1869. Like its still-functioning contemporary counterpart, established the same year next to Asbury Park in Ocean Grove, New Jersey, the Merrick Meeting attracted from 300 to as many as 10,000 attendees to ten-day-long summer prayer assemblies. Although the Merrick Meeting disbanded during the 1920s, its distinctive circular road pattern ringed by tiny cottages built on only slightly larger close-set house lots survives. Locals continue to refer to this neighborhood located just north of present-day Camp Avenue as the

Campgrounds or Tiny Town. Today, Merrick continues to adorn its namesake municipalities of Merrick and North Merrick, the waterways of Merrick Creek and Merrick Bay, the twenty-five-acre Meroke Preserve managed by the Nassau County Department of Parks, Recreation, and Museums, the Merrick Road Town Park and Golf Course, and the two Merrick high roads that meet at the crossroads in the heart of the village of Merrick. The north–south Merrick Avenue runs from Westbury to the south shore at East Bay. The east–west road that crosses it at the Merrick village center starts as Merrick Boulevard in Jamaica, Queens, and runs through Nassau County as Merrick Road before becoming the Montauk Highway after it crosses into Suffolk County.

METEDECONK (Monmouth and Ocean counties, New Jersey). Whritenour (in Boyd 2005:445) wrote that Metedeconk might be a Northern Unami word meaning "at the bad river," from *met*, "bad," *hittuk*, "river," followed by a locative suffix. Rementer in the same source proposed *mahteenekung*, "place of rough ground." Today, Metedeconk is the name of a major river system whose northern and southern branches rise just one mile from each other along the divide separating the inner and outer coastal plains south of the village of Smithburg in Monmouth County's Freehold Township. The twenty-mile-long North Branch passes through the 3,932-acre Turkey Swamp Wildlife Management Area and the 1,180-acre Turkey County Park in Freehold Township to form the boundary between Monmouth and Ocean counties until it reaches a point a little less than two miles above its junction with the South Branch at Forge Pond. The combined Metedeconk River then flows through Ocean County's Brick Township into the mile-wide Metedeconk Inlet where it passes the hamlet of Metedeconk. This hamlet is located on a narrow peninsula lying along the north bank of the lower estuary just across from Metedeconk Neck.

The name Metedeconk first appeared as a locality called Memeocameke in an Indian deed signed by the sachems of Ramenesing on June 18, 1675 (New Jersey Archives, Liber 1:290[49]–289[50]). It was next mentioned as the name of a river called Matedecunk in a deed to land near its mouth dated September 3, 1694 (Whitehead et al. 1880–1931[21]:234). The name then appeared as a tract called Maticonke in a February 12, 1698, Indian deed to land around the Metedeconk River's headwaters (New Jersey Archives, Liber AAA:69). Never removed from local maps, the name has bounced around a bit among the municipalities lining its shores during the past two centuries. Residents of the early nineteenth-century postal village of Metecunk at the head of the inlet changed their community's name to Burrsville before settling on the current name, Laurelton. Today, the little hamlet on the north shore of the Metedeconk's tidal inlet just south of the borough of Point Pleasant continues to bear the river's name. The Metedeconk National Golf Club and the seventy-six-acre Metedeconk Preserve managed by the New Jersey Conservation Foundation both lie along the South Branch of the river. A local fourth-grade class winning the right to select the new preserve's name chose Metedeconk in 2007.

METTACAHONTS (Ulster County, New York). Mettacahonts Creek is a small stream that flows into Rochester Creek, an upstream tributary of Rondout Creek, at the unincorporated hamlet of Mettacahonts in the town of Rochester. The name first appeared as Magtigkenigkonk, one of several small creeks in the area mentioned in the November 2, 1708, Beekman deed (O'Callaghan 1864:89–90). Local residents often used various spellings of Mattekhonk Kill when referring to present-day Rochester Creek. The hamlet of Mettacahonts grew up during the nineteenth century a few miles north of the stream's junction with Rondout Creek at the village of Accord. Located along a wagon road much used by tanners and tradesmen going up to Samsonville and the other small communities scattered below the southeastern slopes of the Catskill Mountains, the hamlet became large enough to support a post office by the early 1880s (Kaiser 1965). Today, Mettacahonts remains on maps as the name of an upper branch of Rochester Creek; the steep, deep gorge along its course; the aforementioned hamlet; and a surviving part of the Old Mettacahonts Road running just north of the village.

METUCHEN (Middlesex County, New Jersey). Nora Thomson Dean thought Metuchen's earliest known recorded spelling, Matockshoning, sounded like a Southern Unami word, *Mahtaks'haning*, "prickly pear creek." Whritenour suggests that it sounds more like a Munsee word, **mahtaaksunung*, "place of the prickly pear cactus." Today, Metuchen is the name of a borough first incorporated in 1900. The September 14, 1677, Indian deed containing the earliest known version of the name identified it as a place where a stake marked the boundary between the towns of Woodbridge and Piscataway (New Jersey Archives, Liber 1:251[89]–250[90]). The name was variously spelled Matuchen, Mettuchinge, and Metuchin in later colonial records. News reports documenting the diplomatic efforts of a Mahican man named Metoxen in frontier politics during the early national period may account for the local tradition holding that Metuchen was the name of a central New Jersey sachem.

MIANUS (Fairfield County, Connecticut, and Westchester County, New York). Today, Mianus is the name of a river, a reservoir, a municipality, and much else along the border between New York's Westchester County and Fairfield County in Connecticut. It is most notably the name of the twenty-one-mile-long Mianus River, which flows from its headwaters in the town of North Castle, New York, through neighboring Bedford, where it crosses the state line into the town of Greenwich, Connecticut. From there the river enters Stamford, where it flows into the Long Island Sound at Stamford Harbor. Land along most of the Mianus River's course is managed at one point or another by an assortment of public, nonprofit, and not-for-profit agencies. The name first appeared in a flurry of Indian deeds signed between 1700 and 1702 conveying land around the headwaters of the river (in Robert Bolton 1881[1]:29, 31). One of these, a 1701 conveyance backdated to 1686, referred to the Mianus River as a stream called Rechkawes or Kechkawes (Gehring 1980:62–63). The relatively late appearance of the stream's current name in colonial records does nothing to detract from local traditions holding that it was named for the Indian warrior Mayane, who single-handedly attacked three Dutch soldiers, killing two before being slain by the third during Governor Kieft's War (Anonymous 1909:281). Examples of the place name Mianus on current maps include the river itself, the North Mianus and Mianus neighborhoods of Stamford at its mouth, the privately owned Mianus River Gorge Preserve, the publicly managed Mianus River State Park Scenic Preserve, the Mianus River Park, several streets and roads, and the Mianus River Railroad Bridge (also known as the Cos Cob Bridge) listed in the National Register of Historic Places in 1987. The name also appears in the form of Myanos, a road in the town of New Canaan, Connecticut, whose spelling clearly indicates that it was meant to celebrate the long-dead Indian warrior. Several locales formerly bearing the name Mianus have been given new names. The Mianus Reservoir and Dam, for example, were renamed to honor the memory of Connecticut politician Samuel L. Bargh. More recently, Interstate 95's Mianus River Bridge was renamed for Michael L. Morano, one of the people killed when the span collapsed into the river in 1983.

MINDOWASKIN (Union County, New Jersey). Mindowaskin Park and Pond are located in Westfield Township. Both were created by township officials during the early 1900s as municipal recreational facilities. The name is a slightly revised spelling of Mindowashen, one of the four Indians mentioned in the deed conveying land in the area to colonists on October 12, 1684 (New Jersey Archives, Liber A:262). He was perhaps also documented under the name Mindawassa, a local sachem and war captain who served his people as a soldier, diplomat, and agent representing Indians selling land in northern New Jersey between 1657 and 1690. The thirteen-acre park is presently maintained by the Westfield Recreation Commission with the help of the Friends of Mindowaskin Park.

MINETTA. See **MANETTO**

MINGAMAHONE (Monmouth County, New Jersey). Whritenour (in Boyd 2005:446) suggests that this name means "big salt lick," from the Delaware words *mengi*, "big," and *mahoni*, "salt lick." Today, eleven-mile-long Mingamahone Brook rises in the Hominy Hills in Naval Weapons Station Earle. From there it flows south to its junction with the Manasquan River between the villages of Squankum and Lower Squankum. It appears in the form Mingumehone Branch as early as the February 23, 1801, act of incorporation of Howell Township (Bloomfield 1811:17–18. It is subsequently mentioned with increasing frequency in late nineteenth-century and early twentieth-century geology reports as Mingumhone or Mingamahone Brook.

MINISCEONGO (Orange and Rockland counties, New York). Minisceongo Creek and its branches drain a substantial area along the easternmost slopes of the Ramapo Mountain Ridge. The name first appeared as Menisiakoungue Creek in an April 16, 1671, patent to land at Haverstraw (Whitehead et al. 1880–1931[21]:17). The South Branch of the Minisceongo Creek rises at the hamlet of New Hempstead just across the low divide that separates it from the headwaters of Pascack Brook. The South Branch flows north past the Minisceongo Golf Course in Pomona to its junction with the Main Branch of the Minisceongo Creek just east of Cheesecote Mountain Town Park. The Main Branch flows from its headwaters around Harriman State Park's Lake Welch east through the town and city of Haverstraw. It then turns north as it flows through West Haverstraw to the place where it joins with Cedar Pond Brook at Stony Point just before their waters flow together into the Hudson River at the Minisceongo State Tidal Wetlands. The Tiorati Brook tributary of the Minisceongo Creek rises at Lake Tiorati just across the Orange County line. From there it flows east into Rockland where it joins Cedar Pond Brook at Cedar Flats just north of Pyngyp Hill on its way to its junction with the Minisceongo Creek at Stony Point. A map drafted by John De Noyelles in 1759, showing patent boundaries claimed by local settlers, identified the present-day Minisceongo creek drainage system as Cheesecocks Brook and its north and south branches.

MINISINK (Sussex County, New Jersey; Orange and Sullivan counties, New York; Monroe County, Pennsylvania). Heckewelder thought the word he spelled Menesink referred to "the habitation of the Minsi tribe of Delawares." Whritenour writes that the name is a virtual dead ringer for a Munsee word, **munusung*, "at the island or place of islands." Goddard (2010:278n3) suggests that the Munsee form *më' n'siiw* and the Unami cognate *mwë'nssi* are variants of old Delaware words for "island." He translates Minisink as a Munsee word, *mënë' sënk,* "at or on the mënë's." Benjamin Barton (1798:2) cited information sent to him by Heckewelder noting that Munsees he knew thought their name derived from the word *monnisi,* a term referring to a long part of an island or a peninsula. Heckewelder added that many Munsees believed their ancestors originally "lived in or under a lake." Lion Miles (2006:personal communication) observed that the root of the Dutch word *bachom* in the inscription "Minnesinck ofte de landt van Bachom" (Minnesinck or the Land of Bachom), on various versions of the Jansson-Visscher map of the region produced between 1650 and 1777, could mean "basin or artificial lake." He found local historian Samuel Eager's (1846) suggestion that Minisink meant "land from where the water had gone" to Indians who thought that an ancient cataclysmic breach at the Delaware Water Gap drained an ancient lake above the Kittatinys entirely plausible. Whritenour finds little merit in Eager's conjectures.

Today, Minisink is a favored name much used for parks, streets, and other places throughout the Greater New York area. It most prominently pops up in online queries and modern maps as the town name of Minisink and much else in the Minisink Valley in New York's Orange County. The word also shows up as the name of the Minisink Hills community across Interstate 80 from the village of Delaware Water Gap in Monroe County, Pennsylvania, and as Lake Minausin originally constructed in 1902 as a focal point for the Pocono Manor Inn in the Monroe County village of Pocono Manor. It further occurs as the sprawling Minisink County Park just east of Lake Hopatcong, New Jersey, and as the name of the village of Minisink Ford and the nearby Minisink Battlefield Park, which marks the place where Indian and Tory loyalists led by the Mohawk captain Joseph Brant overcame the Orange County militiamen sent out against them on July 22, 1779 (Hendrickson, Inners, and Osborne 2010).

It is perhaps not all that surprising that this list does not include the place most intimately connected with the name. The quiet backwater long called Minisink Island that lies along a narrow stretch of lowlands along the Upper Delaware River in present-day Montague Township contains considerable archaeological evidence showing that the area had been a major center of Indian settlement since at least the beginning of Late Woodland times around one thousand years ago. It retained this status throughout the 1600s and well into the following century. Indians at Minisink pressed by colonists continually trying to survey and occupy lands they denied selling held on until outflanked by the loss of the Walking Purchase lands across the river after 1737. Most gradually moved farther westward to refugee towns on the other side of the Poconos and Catskills along the upper reaches of the Susquehanna River between Wyoming and Oquaga. Their determined efforts to drive colonists from the area during the final French and Indian conflict and throughout the Revolutionary War failed. Those trying to return to their homes in the valley after the fighting stopped found themselves terrorized by indiscriminately murderous "Indian killers." Most finally gave up and moved farther west. Their name moved west as well, gracing such communities as Minnisung, Wisconsin, and Munising, Michigan. Today, the Minisink Archaeological District, designated as a National Historic Landmark in 1993, commemorates the significant role Indian people in Greater New York played in the early history of America. Unmarked and undeveloped, Minisink Island continues to lie quietly unnoticed by canoeists and motorists passing by along the upper end of the Delaware Water Gap National Recreation Area.

MINNEAKONING (Hunterdon County, New Jersey). Minneakoning is the name of a one-mile-long creek and a road that runs alongside it in the borough of Flemington. Rising near the Hunterdon County Medical Center, Minneakoning Creek flows almost due east to its junction with the South Branch of the Raritan River. It first appeared as a run or brook spelled Majs-Wonawayskowong and Mawaonawoghkonong in an Indian deed to land in the area dated June 5, 1703 (New Jersey Archives, Liber AAA:443–44). It next appeared as a brook called Mineaukoning that formed the southern boundary of an adjoining tract sold on October 7, 1709 (ibid., Liber BB:323–24). The close correspondence in the spellings of the latter and current place name suggest that both share a common origin in the 1709 deed.

MINSI. See MUNSEE

MOHAWK (Delaware and Schoharie counties, New York). It is not unusual for nations to be known by names given by not-always-well-disposed neighbors. Mohawk is one of these; a Southern Unami version is *mhuweyok*, "cannibal monsters." Members of this easternmost Iroquois nation, whose name itself came from either a Huron word for "real adder" or a Basque expression for "killer," called themselves *Kanyékeha:ka*, "people of the place of the flint." Mohawk is also about the only known Iroquois word known to have been used as a place name in present-day Greater New York during colonial times. It particularly applied to the Mohawk Branch of the Delaware River, whose waters rise around the 3,209-foot-high Mount Utsayantha at the northwestern end of the Catskill Mountains. The name made an early appearance on maps depicting the upper reaches of the Delaware River as the Mohock Branch in William Scull's 1770 survey map. Today, Scull's Mohock Branch is called the West

Branch of the Delaware. The range of hills separating the valleys of the Mohawk Branch of the Delaware from the North Branch of the Susquehanna represented the border between the Munsee and Mohawk homelands during the eighteenth century. The Mohawk Branch itself flows from its headwaters in Schoharie County just north of Mount Utsayantha into Delaware County, where it drains the western slopes of the Catskill range from Stamford through Delhi and Walton into the Cannonville Reservoir. From there it flows past Deposit to Hancock, where it joins with the East Branch to form the main stem of the Delaware River. Contending colonists exploited claims made on the area by Mohawks and Munsees to press those of their own. Colonial authorities reached an accord establishing a mutually agreed boundary between Indian and colonial lands at the Fort Stanwix Treaty in 1768. A private purchase of land between the Pepacton and Mohawk branches made at the meeting continued to roil the waters. Lawyers trying to gain advantage for one contender or another continued to fight it out in court chambers and boardrooms until a special commission set up by New York State settled things by awarding the 1768 buyers equivalent lands west of the Mohawk Branch in 1786.

Mohawk people played an altogether different role in more recent history in the Greater New York area when many, mostly from Canadian reserves near Montreal, found employment in Manhattan as ironworkers. Working on bridges and skyscrapers, they rented apartments in the Gowanus section of Brooklyn. Writer Joseph Mitchell (1949) tellingly described life in the small Mohawk community that grew up there during the early decades of the twentieth century. Their attendance at Indian get-togethers at Inwood Park also helped establish associations linking Indians to the northernmost cliffs and waterways at the northern end of Manhattan in the minds of many New Yorkers. Other examples of the place name Mohawk in the region are discussed in Part 2.

MOHINGSON (Monmouth County, New Jersey). Rementer (in Boyd 2005:425) suggested that Mohingson sounded like a Southern Unami word, *mhuwingwsink*, "place of the blackberries." Mohingson Brook is a four-mile-long stream that rises along the northern slope of the Mount Pleasant Hills in Holmdel Township. Flowing northward, the brook passes into Aberdeen Township through the borough of Matawan and into the wetlands, where it mixes with Matawan Creek waters flowing into Raritan Bay. Mowhingsinnge Creek was first mentioned in a local deed confirmation made on May 12, 1688 (Whitehead et al. 1880–1931[21]:119). An Indian deed to land in the area signed on October 10, 1690, mentioned the Mohingsunge tract, neck, and creek (Monmouth County Records, Deed Book C:163).

MOHONK (Ulster County, New York). David Oestreicher (in Fried 2005:22) wrote that Mohonk was a shortened form of a Munsee word, *maxkawenge*, "hill of bears." Whritenour thinks Mohonk sounds more like the Munsee words *maxkwung*, "place of bears," and *mohkwung*, "place of blood." Today, Mohonk is the name of a lake, a resort, and a preserve at the northern end of the Shawangunk Mountain Ridge. The name first appeared as "the high hill Moggonck" in the Indian deed obtained by the Huguenot founders of New Paltz on May 26, 1677 (O'Callaghan 1864:114). Fried (2005:17–23) has shown how local landowners jockeying for advantage variously applied the name to a creek, a pond, and High Top, the promontory that looms above both features. Mohonk in its current form was first bestowed on the mountain lake that currently bears its name by the operator of a resort built on its shores in 1860. Brothers Alfred H. and Alfred K. Smiley retained the name for the Mohonk Mountain House Resort they opened there in 1869. The Smiley family continues to operate the resort, designated as a National Historic Landmark in 1986, and directs management of the surrounding 6,400-acre Mohonk Preserve as the nation's largest not-for-profit, privately owned nature sanctuary.

MOMBACCUS (Ulster County, New York). Mombaccus is another example of an Indian place with a European name. Fried (2005:75–76) has shown that the name first appeared as Mumbackers in three survey descriptions of land at and around the present-day village of Mombaccus in the town of Rochester entered in Ulster County records on September 4, 1676. Mumbacker (literally "mask face") is the Dutch version of the Greek word Mombaccus, the "mask of Bacchus" worn in antiquity during cult rites honoring the god. The first part of the word entered English during medieval times as mummer. In Ulster County, Mombaccus was meant to mark the location of an Indian carving of a man's face on a sycamore tree near the junction of "the Mombakkus and Rondout kills," noted in 1676 (Ruttenber 1906a:169–70; Fried 2005:76). Colonists gave the name to the settlement they built there and the creek that flowed through it. Although both names were officially changed to Rochester when the community joined others to form that town in 1703, residents continue to refer to the original hamlet and a side branch of Mettacahonts Creek as Mombaccus. Mombaccus Mountain, a 2,840-foot-high Catskill peak several miles west of the creek, evidently received its name sometime during the turn of the twentieth century.

MONGAUP (Delaware and Sullivan counties, New York). Whritenour suggests that Mongepughka, an early orthography of Mongaup, sounds like a Munsee word, **mangaapoxkw*, "big rock." Today, places bearing the name cluster in two separate parts of the Catskills. The Mongaup River drains a substantial part of southern Sullivan County. Farther north, the Mongaup Creek flows through the Upper Catskill highlands in Delaware County.

New Jersey surveyor John Reading, Jr. (1915:102, 108), first documented the name of the present-day Mongaup River during the summer of 1719 while working with New York commissioners to establish a mutually acceptable border between their provinces. He noted it first as a place called Mingepughkin on July 8 during his outward journey and as Mongepughka on his return on July 24. Sawmills built beside the many branches of the Mongaup River helped turn the region into a major center for the logging and tanning industry during the early nineteenth century. Impoundments whose waters powered the sawmills have since been converted from millponds to resort lakes and reservoirs that have made the Mongaup one of the hardest-working rivers in Greater New York. The headwaters of the uppermost of these streams, known today as the Middle Mongaup River, rise above the village of Liberty before flowing south to a junction with East Mongaup River water from Hurleyville and Grossinger's Lake. From there the river flows to its junction with the West Branch of the Mongaup River coming in from White Sulphur Springs and Swan Lake. It then flows through the village of Mongaup Valley into the Swinging Bridge Reservoir, where it receives waters from White Lake. The stream flows into the Mongaup Falls Reservoir, where the Black Lake Creek adds water from Lake Superior, Black Lake, and the Toronto Reservoir. Its waters then pass through dam sluiceways at Mongaup Falls into the five-mile-long Rio Reservoir, where waters from Saint Joseph and Sackett lakes pour into it from Black Brook. River water passing through Rio Dam sluice gates roars through the steep gorge cut by the rapidly running stream to the Mongaup's confluence with the Delaware River at the old Delaware and Hudson Canal village of Mongaup.

The name of Mongaup Creek far to the north of the Mongaup River is an artifact created by one of the rival syndicates that fought over Hardenbergh Patent boundaries in 1785. These men asserted that the Mongaup River mentioned as a key patent boundary actually was the much more northerly creek that encompased a significantly larger tract of land. Although their effort did not meet with success, the name Mongaup stuck to the creek that still flows from Mongaup Pond through the Mongaup State Campground and the 11,967-acre Mongaup Valley State Wildlife Management Area into Willowemoc Creek and on to the Beaver Kill into the East Branch of the Delaware River in Delaware County.

MONOCACY (Northampton County, Pennsylvania). Heckewelder thought Monocacy sounded similar to the Delaware words *menágassi* or *menákessi*, "a stream containing several large bends." The ten-mile-long Monocacy Creek in the Lehigh Valley first appeared as Menacasy Creek in the 1738 Evans map of the Walking Purchase (in Gipson 1939). Manakisy Hill, today's Camels Hump, initially was depicted on the 1770 William Scull map. Standardized as Monocacy by the mid-nineteenth century, the name may be native to the area. It could also be a transplant, however, one that traveled north into the Lehigh Valley during the early 1700s from the Monocacy River in Maryland's Potomac Valley via Monocacy Creek in the Schuylkill Valley where Daniel Boone was born, in the more southerly Pennsylvania county of Berks.

MOODNA (Orange County, New York). Moodna comes from the Dutch word *Moordenaar*, "murderer." Tradition holds that the name marks a murderous encounter with local Indians said to have occurred during Governor Kieft's War. Colonists subsequently referred to Native people living along its banks as Murderer's Creek Indians. Sixteen-mile-long Moodna Creek drains a substantial watershed southwest of the city of Newburgh. The creek starts at the junction of Satterly and Otter creeks outside the hamlet of Washingtonville in the town of Blooming Grove. From there it flows west into the town of Cornwall past Salisbury Mills and beneath the Moodna Railroad Viaduct through Vail's Gate to its junction with the Hudson River at Cornwall-on-Hudson.

MOONACHIE (Bergen County, New Jersey). Whritenour holds that Moonachie sounds exactly like a Munsee word, **moonahkuy*, "dug-up land." He also thinks that earlier forms of the name, such as Minckacque, sound like a Munsee word, **meengahkwuw*, "there are great trees." Today, Moonachie is the name of a borough and a creek in the Hackensack Meadowlands. The place was first mentioned as a tract called Minckacque in a boundary confirmation made on October 26, 1661 (Winfield 1872:7–8). Subsequently noted for the quality of its farmland, the locality was known during the early 1800s as Peach Island before reverting to its original name in the forms of Monachie and Moonachie (Van Valen 1900:358–59). Residents incorporated their community there as the borough of Moonachie in 1910. Moonachie Creek takes its rise in the wetlands at the south end of the borough. From there it flows through the marshes in the neighboring borough of Carlstadt to its junction with the Hackensack River just north of Secaucus.

MOONHAW (Ulster County, New York). Moonhaw is the personal name of an Esopus Indian leader. Today, Moonhaw Road and what locals call the Moonhaw Hollow (Maltby Hollow on the few maps that give a name for the place) are modern place names in the hamlet of West Shokan in the town of

Olive. Both the road and the hollow bear the name of an Indian identified as "Moonhaw alias Ancrop" among the many Native signatories who placed their marks on the June 6, 1746, Indian deed to land in the area (Ulster County Records, Deed Book EE:63–65).

MOOSIC (Lackawanna and Wayne counties, Pennsylvania). Whritenour suggests that the present-day spelling of Moosic resembles a Munsee word, **moosak*, "elk" (plural) or **(eenda) moosiik*, "where there are elk." Moosic Mountain lies at the westernmost end of the Munsee homeland. It is a narrow twenty-four-mile-long range of hills that extends from the southwestern corner of Wayne County into the adjoining county of Lackawanna. Moosic hills rise an average of two thousand feet above sea level to a high point of 2,300 feet at Pocono Plateau. The ridge forms part of the divide between the Susquehanna watershed and the uppermost western headwaters of the Lehigh Valley. It was first mentioned in colonial documents as the "large mountain called Moshooetoo mount or hill." This hill served as the western boundary of the New England purchase of land in the area made on May 6, 1755, as well as the frontier separating the lands of the "Ninneepaaues" (Delawares) and the "Macquas . . . otherwise called Mowhawks" who sold the land (Boyd and Taylor 1930–71(1):260–71). The name also appears on present-day maps as the community of Moosic Lakes in Jefferson Township in Lackawanna County.

MOPUS (Fairfield County, Connecticut, and Westchester County, New York). Mopus is currently the name of an upper tributary of the Titicus River that straddles the state line between the town of North Salem in Westchester County and the town of Ridgefield in Fairfield County. It also appears on maps and road signs as the name of a road and a bridge that cross Mopus Brook in Ridgefield. The name first appeared as Mopoos Ridge in an Indian deed to land in the area dated November 22, 1721 (Teller 1878:23–24), and has served as the name of Mopus Brook, Bridge, and Road ever since.

MOUNT KISCO. See **KISCO**

MUCHATTOES (Orange County, New York). Muchattoes Lake and Dam lie within the Quassaick Creek watershed just south of the junction of New York State Routes 17K and 32 in the city of Newburgh. Both the lake and the dam were built by the city in 1912. The name was probably drawn from the entry listed as Much-Hattoos in Newburgh historian Edward Ruttenber's (1906a:129) Indian place name book. Ruttenber found the name in a 1709 petition requesting permission to purchase a tract of land at New

Windsor and a repetition of the name in the 1712 patent officially recognizing the purchase. Both documents noted that the tract was bounded on the west "by the hill called Much-Hattoos." Ruttenber went on to note that local people subsequently gave the name Snake Hill to the Much-Hattoos promontory located southwest of the present-day pond.

MULHOCKAWAY (Hunterdon County, New Jersey). Mulhockaway Creek is a five-mile-long stream that rises above the headwaters of Hakihokake Creek just west of the village of Pattenburg (named for a now defunct local patent distillery) in Union Township. It flows east from there past the sixteen-acre Mulhockaway Creek Preserve managed by the New Jersey Natural Lands Trust into the Spruce Run Reservoir completed in 1963. Reservoir workers direct released impoundment water into artificial channels that run past the Spruce Run Dam into the South Branch of the Raritan River at the freestanding town of Clinton. John Reading, Jr. (1915:44–45), made the first references to Mulhockaway in two journal entries written during the early summer of 1715. The first noted an "Indian path which leads to Monsalockquake at an old Indian plantation called Pelouesse" on May 30. Two days later, Reading identified "Mensalockauke an Indian plantation on the head of a part of a southerly branch of Rarington River" where he spent the night before traveling on the next day to "Essakauqueamenshehikkon an Indian plantation on the back of the Great Swamp." Local residents called the stream Mensalaughaway Creek well into the twentieth century.

MUNSEE (Rockland County, New York, and Northampton County, Pennsylvania). One of the more widespread Indian place names in Greater New York, Munsee is also one of the harder names to place in this book. As mentioned earlier, Goddard's (2010:278) suggestion that Munsee means "person of *mën'ë's*'" (i.e., Minisink Island) is almost certainly correct. His assertion that "'Munsee' was originally just the name of the band headquartered on Minisink Island" is not supported in the colonial literature, which contains no known instance of Indian or colonial use of the name Munsee in reference to a band or anything else in present-day Greater New York. Instead, as Goddard goes on to appropriately observe, the name only began to be used after its first occurrence in colonial records in 1727 by exiled speakers of the Munsee dialect of Delaware and their friends and relatives as they gathered together in refugee communities along and beyond the Susquehanna Valley. Despite this fact, few words have more right to be considered native to the area than the current name of its first people and their language. The fact that all occurrences of Munsee in Greater New York are restorations makes them no different from most other Indian names on present-day maps that are also relatively recent toponymic ressurections. Some of these names, such as the many iterations of Monsey on Rockland County maps since the 1840s, predate more modern reintroductions. Others, such as Mount Minsi on the Pennsylvania side of the Delaware Water Gap, Lake Minsi constructed in Northampton County in 1970, and the Minsi Trail Bridge built across what is regarded as an Indian trail ford across the Lehigh River in Bethlehem, are among the various more recent permutations of Munsee attached to municipalities, lakes, roads, camps, mountains, and other places in the region. Long Island's Munsey Park, listed in this book's second section, bears the family name of French origin (sometimes spelled Mouncey) of its developer.

MUSCONETCONG (Hunterdon, Morris, Sussex, and Warren counties, New Jersey). Heckewelder's identification of Musconetcong as a Munsee toponym, *maskhanneccunk*, "rapid running stream," has long been accepted as a most plausible etymology. Revisiting the word, Whritenour finds that it sounds more like **maskaneetkung*, Munsee for "strong, fertile lowland." Today, Musconetcong most prominently occurs as the name of a valley, a mountain, a lake, and the forty-six-mile-long river that forms at one place or another a boundary line between four New Jersey counties. Rising at Lake Hopatcong, the river flows as an upland stream into Lake Musconetcong past the borough of Netcong (a truncated abbreviation of Musconetcong adopted when the borough incorporated in 1894) and the historic village of Waterloo to Hackettstown. Below Hackettstown the river follows a generally straight and narrow course through the Musconetcong Valley between the Pohatcong and Musconetcong mountain ridges as it flows past a series of small river hamlets to its junction with the Delaware River at Riegelsville. Like many other names in northwestern New Jersey, the Musconetcong River was first mentioned in Indian deeds and surveys determining their bounds undertaken during the second decade of the eighteenth century. It appeared under several spellings in the four West Jersey Society Indian deeds to lands around the Delaware Water Gap signed between January 17, 1712, and August 18, 1713 (New Jersey Archives, Liber BBB:140–47; Liber GG:458–60). John Reading, Jr. (1915), spelled the name of the river Muskonetkong in the many references he made to the stream while measuring tracts allotted to West Jersey Society investors between 1715 and 1719. In use ever since, the name still adorns the river as well as parks, preserves, roads, and the range of hills paralleling the river between Schooleys Mountain at its northern end and Mount Joy, located where it flows into the Delaware River.

MUSCOOT (Putnam and Westchester counties, New York). Muscoot is the name of a river that rises in Putnam County and flows south past a park on the shores of a reservoir that both bear the name in Westchester. Muscoot first appeared in colonial records in a reference to a river called Moscotah mentioned in a May 5, 1703, Indian deed to land in the area (Marshall et al. 1962–78[4]:404). Today, the Muscoot River flows south from Lake Mahopac into the Amawalk Reservoir. Water released through Amawalk Reservoir Dam sluiceways flows southeast to the river's now-flooded junction with the Croton River at Muscoot Reservoir. Completed by the City of New York in 1905, the Muscoot Reservoir collects all Croton Valley waters held back behind its dam. Muscoot Farm, located above the Muscoot Reservoir high-water mark where its waters flow into the Croton River, was the summer estate and working dairy farm of the Hopkins family from 1880 to 1924. Acquiring the 777-acre tract in 1967, the Westchester County Department of Parks, Recreation, and Conservation maintains the property as a historic farm celebrating the county's agrarian heritage.

MUSCOTA (New York County, New York). Muscota New School P.S. 314 in the Inwood section of northern Manhattan bears the Indian name of the Harlem River. The name was first recorded as the Kil Muskota in a deed confirming Adriaen van der Donck's 1646 purchase of land in Riverdale and Yonkers (since lost) executed on September 11, 1666 (Palstits 1910[1]:234–35).

MUSQUAPSINK (Bergen County, New Jersey). Musquapsink Brook is a five-mile-long stream that flows into the Pascack Brook at the borough of Westwood. Harvey (1900:35–36) wrote that he had seen references to Musquampsont Brook in papers associated with early purchases of land first patented to Samuel Bayard in 1703. The stream was subsequently noted in an August 15, 1753, deed as Masquamp or Masquap Creek (in Budke 1975b:6). A French military engineer's map drafted around 1781 (in Leiby 1962) noted a rivulet identified as Little Pascack River at Musquapsink Brook's present-day location. Like many other remote or obscure topographical features in New Jersey, Musquapsink Brook began to appear with some regularity in geological reports published around the turn of the twentieth century.

MYANOS. See **MIANUS**

NAACHPUNKT (Passaic County, New Jersey). Naachpunkt Brook is a three-mile-long tributary of Preakness Brook that serves as a part of the northwest border of the borough of Totowa. The junction of Naachpunkt and

Preakness brooks forms the Singac Brook, which flows south alongside Totowa into the Passaic River. William Nelson (1902:188–89) provided a unique first-person account describing his role in selecting the particular form of this place name for adoption in the official records. Retained in 1898 to help draft an act of the state legislature to erect Totowa into a borough, Nelson thought that the place name spelled Nackpunck then attached to the creek came from a Dutch word, *naaktpunkt,* meaning "bare point." Nelson's subsequent talks with local old-timers who remembered it as an Indian name led him to deeds dated between 1686 and 1709 that contained the marks of an Indian signatory's name variously spelled Nackpunck, Machpunk, Moghopuck, and Mackapoekat. Despite these findings, his Dutch etymological form of the name remains on present-day maps.

NAKOMA (Rockland County, New York). Nakoma Brook is a small stream located in the town of Ramapo. Nakama was the name of one of the Indians who signed the deed to a tract of land identified as Pothat (also spelled Potake) on March 7, 1738 (Cole 1884:264). Local residents finding Nakama in a copy of the deed probably gave it to the creek in the slightly revised form that more closely resembled the well-known Indian name Nokomis from Longfellow's *Hiawatha.*

NAMANOCK (Sussex County, New York, and Monroe County, Pennsylvania). Whritenour suggests that Namanock sounds much like a Munsee expression, *ne meenaxk,* "that fort." Namanock is probably a variant spelling of Nowenock, a sachem whose name appears elsewhere in the region at Nanuet and several other locales. Today, Namanock is the name of an island located on the New Jersey side of the Delaware River one mile south of Minisink Island in Sandyston Township. Ruttenber (1906a:222) stated that Reverend Casperus Freymout referred to the place as "an island so called" in 1737. Colonists built Fort Namanock nearby at the beginning of the last French and Indian War in 1756 as part of the defense line of fortified houses and small stockades that stretched along Pennsylvania's Blue Ridge up along the Delaware River past Port Jervis to the New York provincial frontier. The ruins of Fort Namanock stood on the New Jersey shore just across from Namanock Island until the 1960s, when they fell prey to demolition teams clearing land for the since-cancelled Tocks Island Dam and Reservoir. Today, Namanock Island is located within the borders of the Delaware Water Gap National Recreation Area. Two other places in the general vicinity also currently bear the name. The hamlet of Normanook five miles southeast of Namanock Island was the location of the Forest Service's now-demolished

Normanook Fire Tower. Namanock Trail is a street in the village of Pocono Lake, Pennsylvania, several miles to the northwest of its namesake island.

NANNAHAGAN (Westchester County, New York). Nannahagan Brook, Park, and Pond and Nanny Hegan Road are located in the village of Thornwood in the town of Mount Pleasant. The name evidently first appeared when Huguenot immigrant Isaac Sie and his four sons settled at a place called Nanageeken or Nanequeeken in upper Westchester around 1690 (Shea 1999). The anglicized version of the name presently embossed on Nanny Hagen Road signs has been on area maps since at least 1778. Nanny Hagen Road splits from Kensico Road just east of Thornwood. From there it runs east three miles to its junction with New York State Route 120 just above the northern end of Kensico Reservoir. Nannahagan Brook rises just west of the latter junction, flowing in a gentle arc into Nannahagan Pond and Park to its junction with the Saw Mill River at Thornwood.

NANUET (Rockland County, New York). Whritenour thinks Nanuet sounds somewhat like a Munsee name, *neenawiit,* "he who recognizes me." Today, Nanuet is the name of a hamlet in the town of Clarkstown. Local tradition holds that Nanawitt's Meadow, the first name settlers gave to the area, referred to an Indian signatory mentioned by that name in the March 5, 1703, Wawayanda Patent deed (Budke 1975a:79–81). New York and Erie Railroad officials renamed the place Clarkstown Station when they ran the line through the locale in 1841. Village residents subsequently chose the name Nanuet for the post office that opened there in 1856. It first appeared as Manuet, the spelling of the name of an Indian recalled by a local farmer making a deposition in 1785 regarding Indian deed boundaries in the area (Hommedieu 1915:13). Together, Nanawitt and Nanuet are probably two spellings of Nowenock, the name of the aforementioned Indian leader who played a major role in many land sales along the highlands between the Pequonnock country in westernmost Connecticut and the lands along the New York–New Jersey border from 1696 to 1727.

NAPANOCH (Ulster County, New York). Napanoch is the name of a village located at the place where Rondout Creek joins with Sandburg Creek, at the base of the Shawangunk Ridge in the town of Wawarsing. The name first appeared in colonial records as Nepenaack in the Indian deed to land in the area dated June 8, 1696 (Ulster County Records, Deed Book CC:145). A group of Native people identified as Nappaner Indians attended a Nicolls Treaty renewal meeting held on June 9 and 10, 1719 (New York State Library, Executive Council Minutes [11]:607–13). Nepenagh was noted in an

Indian deed conveying land containing a lead mine in June 15, 1728 (Ulster County Records, Deed Book DD:6–7). The name was also given to a mill built alongside Rondout Creek in 1754 and to the fort erected nearby in 1756 as a link in the region's colonial frontier defense line. The Delaware and Hudson Canal built through Napanoch in 1828 helped make the locale a transportation and manufacturing center. By 1844 the community was large enough to support a post office (Kaiser 1965). Much of Napanoch subsequently moved next to the station built by the Delaware and Hudson Railroad, a mile east of the canal. The Eastern States Correctional Facility opened at the old canal port at Napanoch in 1900.

NARANEKA (Fairfield County, Connecticut). Naraneka Lake and Dam are located in the town of Ridgefield. Completed in 1938, both the lake and the dam were originally named for local landowner Seth Low Pierrepont. Pierrepont subsequently renamed the pond Lake Naraneka, a slightly altered spelling of the name of the sachem who signed the November 22, 1721, Indian deed to land in the area as "Tackore otherwise called Norreneca" (Teller 1878:23–24). Like Namanock and Nanuet, both Naraneka and nearby Norrans Ridge are evidently variant spellings of Nowenock. Other places bearing different names borne by this sachem on present-day Ridgefield maps include Oreneca Road and Tackora Trail.

NARRASKETUCK (Nassau and Suffolk counties, New York). Whritenour thinks Warrasketuck, the earliest known spelling of Narrasketuck, sounds like a Munsee word, *wulaskihtukw*, "river of good pasturage." Today, Narrasketuck Creek forms the border between the Nassau County village of Massapequa and Amityville in Suffolk County. The name first noted as Warrasketuck identified a creek at the western end of a section of Massapeague Meadows sold to settlers in an Indian deed dated March 17, 1658 (Cox 1916–40[1]:347–49). Twenty years later it was identified as Narrasketuck in the patent Governor Andros granted for land in the area on September 29, 1677, to the town of Oyster Bay. On local maps in that form since that time, Narrasketuck was adopted as the name of the local yacht club when it was founded in 1933.

NATIRAR. See **RARITAN**

NAURAUSHAUN (Rockland County, New York). Nauraushaun Brook and village are located in the town of Orangetown. Rising just above Nanuet, the six-mile-long brook flows past the hamlet of Nauraushaun at the north end of Lake Tappan. The name first appeared on the west bank of the Hudson

River as Narratschone, just north of the Tappaans on a draft copy of John Seller and William Fisher's map of 1677. It subsequently appeared as "land called Narranshaw" in the same approximate location in the Kakiat Patent Indian deed signed on June 25, 1696 (Goshen Public Library, Kakiac Patent Papers). It next appeared in the June 1, 1702, Indian deed to the nearby Narrachong tract (Budke 1975a:84–86) and was later recorded as the name of a stream spelled Narashunk in a May 9, 1710, Indian deed (New Jersey Archives, Liber I:317–19) and as Narranshaw creek or brook in the 1713 survey of the 1710 sale (Green 1886:34). Local residents have referred to the area around the brook as Nauraushaun since the eighteenth century. The modern-day hamlet of Nauraushaun was variously called Sickeltown and Van Houten's Mills at different times during the nineteenth century (Wardell 2009:64).

NAVESINK (Monmouth County, New Jersey). Eastern Oklahoma traditionalist Lucy Parks Blalock (in Boyd 2005) suggested Navesink was a Southern Unami word meaning "place where you can see from afar." Today, Navesink is the name of a tidal river, the community on the river's north shore, and much else at the northeast end of Monmouth County. The name made its first appearance in colonial records on March 28, 1651, when "Indians of Nevesincx" sold land along the south shore of what colonists a year later called Neversinck Bay (New Jersey Archives, Liber 1:7–8; O'Callaghan and Fernow 1853–87[13]:31–32). Colonists from Gravesend repurchased the lands at Navesand on what they now called Raritan Bay from the Indians in the spring of 1664 (Municipal Archives of the City of New York, Gravesend Town Records, Deeds:72–73). Colonists and Indians both often referred to the region in and around present-day Monmouth County as Navesink during the colonial era. An altogether different group of places called Neversink (sometimes confusingly spelled Navesink) is located far to the northwest in the Catskills. Streams belonging to the Navesink watershed in central New Jersey drain a substantial part of the area. Only the eight-mile-long estuary that flows into Sandy Hook Bay, however, is known as the Navesink River. Even here the name is unofficial (the state formally identifies it as the North Shrewsbury River). The small hamlet of Navesink has occupied its current locale since the late 1600s. The Atlantic Highlands just north of the hamlet are also often called the Navesink Hills. In 1841 Navesink Light, today's Twin Lights Historic Site (designated as a National Historic Landmark in 2006), became the first place where unprecedentedly powerful Fresnel lenses were installed. Today, the toponym continues to serve as the name of the Navesink Beach barrier island community as well that of the sixty-five-acre Navesink River Wildlife Management Area and a variety of municipal agencies, local businesses, and streets in the area.

NEGUNTATOGUE (Suffolk County, New York). Neguntatogue Park and Creek are located in the hamlet of Lindenhurst in the town of Babylon. The name first appeared in a sale of meadow land on "the neck called Neguntetake" made on March 2, 1663 (C. Street 1887–89[1]:55). It later appeared as Naggunttatouge Neck in an Indian deed to land in the area signed on July 11, 1689 (ibid.:33–36), and as Neguntategue Neck in a November 20, 1705, Indian deed (ibid.:291–94). Lindenhurst residents named their park after the small one-mile-long Neguntatogue Creek inlet that still flows south from the village into the Great South Bay.

NEPAS (Fairfield County, Connecticut). Nepas Road is located in the Lake Hills development built in 1952 in the town of Fairfield. Furnished at the developer's request by the Fairfield Historical Society, Nepas first appeared as Neesenpaus, the name of one of the Indians who signed the October 6, 1680, Indian deed to land in the area (Wojciechowski 1985:105).

NEPPERHAN (Westchester County, New York). Today, Nepperhan is the name of several streets in and around the Nepperhan Heights and Nepperhan Park neighborhoods in the city of Yonkers. The name first appeared as Neperan in the September 11, 1666, Indian confirmation of the since-lost Van der Donck (the original *jonkheer*, "young lord") deed to land in and around Yonkers (Palstits 1910[1]:234–35). The stream variously identified as Nippizan, Nipperan, Nepperha, Wepperhaem, and Neppierha was mentioned as the Indian name for what colonists called Younker's Creek or Kill in deeds to lands in the area signed between 1681 and 1685 (in Robert Bolton 1881[1]:268–71, 507; [2]:2–3). The many dams built across the Nepperhan Creek at several locales in Yonkers led residents to refer to the stream as the Saw Mill River. These provided the fall of water that powered the many factories built in the area during the mid-nineteenth century. Local residents still refer to the land along the lower course of the Saw Mill River as the Nepperhan Valley.

NESCOPECK (Luzerne County, Pennsylvania). Heckewelder thought Nescopeck sounded like a Delaware word, *nækchöppeek*, "blackish, deep, and still water." Whritenour finds that Nescopeck sounds more like a Munsee word, *niiskpeek*, "that which is wet." Although most places bearing the name Nescopeck lie just outside of the metropolitan area in the Susquehanna

Valley, the southeasternmost slopes of Nescopeck Mountain drain streams that flow into the Upper Lehigh Valley. Nescopeck Mountain is a long ridge that forms another part of the divide separating the western headwaters of the Lehigh River from waters falling into the Susquehanna River. On November 21, 1740, Pennsylvania provincial secretary James Logan noted that Forks Indian leader Nutimus and his people had withdrawn behind what he identified as the "mountain called Neshameck" after being evicted from their lands in the Lehigh Valley (American Philosophical Society, Logan Papers [4]:71–72). Nescopeck Creek, a tributary of the Susquehanna River, first appeared on William Scull's 1770 map. One of its branches, the Little Nescopeck Creek, edges into the westernmost part of the metropolitan area within a mile of the Lehigh River above White Haven.

NESHAMINY (Bucks County, Pennsylvania). Heckewelder thought Neshaminy sounded much like a Delaware word, *neshâmhanne*, "two streams making one." Neshaminy Creek's North and West Branch headwaters flow across the southwesternmost borders of the Munsee homeland. The earliest appearances of the name related to those portions of the creek's drainage within Munsee country primarily occurred in deeds and other land papers involving the Penn family's 1737 Walking Purchase claims.

NESHANIC (Hunterdon and Somerset counties, New Jersey). Nora Thompson Dean thought Neshanic sounded much like a Southern Unami word, *nishanèk*, "two rivers." Neshanic is the name of the ten-mile-long river system whose route traces a great arc from its headwaters in Hunterdon across the Somerset line to its junction with the South Branch of the Raritan River just west of the villages of Neshanic and Neshanic Station. The southernmost headwater of the system, known as the Third Neshanic River, rises at Sand Brook in Delaware Township. From there it flows east and north to its confluence with the eastward-flowing two-mile-long Second Neshanic River in Raritan Township. Together, these streams form the First Neshanic River, which flows below the borough of Flemington into East Amwell Township. From there it crosses the Somerset County line into Hillsborough Township, where the stream is known simply as the Neshanic River.

The name first appeared in colonial records as "a branch of Rariton River called Noshaning" in an Indian deed to land in the area dated November 11, 1703 (New Jersey Archives, Liber AAA:434–35). Colonists began settling in and around the small community that soon became known by the anglicized form New Shannock by the mid-1700s. Regaining its Indian associations as Nashanic by the early 1800s, the hamlet grew up along the Amwell Wagon

Road just above the northernmost end of what were then known as the Nashanic or Rock Mountains (today's Sourland Mountains). The focus of community activity shifted north to the old mill site where twin bridges carried the tracks of the Lehigh Valley Railroad and the Central Railroad of New Jersey across the South Branch of the Raritan at present-day Neshanic Station. Today, the name is also borne by the Neshanic Mills and the Neshanic historic districts listed in the National Register of Historic Places in 1978 and 1979, respectively, as well as the Neshanic Valley Golf Course opened by the Somerset County Park Commission in 2004.

NESHISAKAWICK (Hunterdon County, New Jersey). Whritenour thinks Neshisakawick sounds like a Northern Unami word, *neschi-sakquik,* "double outlet or mouth." Today, Neshisakawick adorns the names of two creeks that flow next to one another at the western end of Hunterdon County. The uppermost branches of the main stem of Neshisakawick Creek rise less than a mile from the headwaters of Cakepoulin Creek at the northeastern end of Alexandria Township. Joining together at the hamlet of Everittstown, Neshisakawick waters wend a winding course past cliffs and ledges along the lower four-mile-long section of the creek to its junction with the Delaware River in the borough of Frenchtown. Little Neshisakawick Creek flows through Kingwood Township to join the Delaware River less than half a mile south of its namesake's outlet. Present-day Neshisakawick Creek was first noted in colonial records as the "brook called Noshasakowick" that served as part of the northern bounds of the aforementioned November 11, 1703, Indian purchase (New Jersey Archives, Liber AAA:434–35). It was later identified as Neshasakaway, forming a part of the southern bounds of land sold in the August 16, 1711, Indian deed (ibid., Liber BBB:206–207).

NESQUEHONING (Carbon and Schuylkill counties, Pennsylvania). Heckewelder translated the name as a Delaware word, *næskahónink,* "at the black lick," from the root word *næskahóni,* "black lick or the lick of which the water is of a blackish color." Today, Nesquehoning Mountain, Creek, and the municipalities of Nesquehoning and Nesquehoning Junction are located in the upper Lehigh Valley. Nesquehoning Mountain is one of the series of long, narrow parallel ridges that follow one after another above the Blue Mountain. The eleven-mile-long Nesquehoning Creek runs along the northern base of the mountain from its headwaters in the Schuylkill County hamlet of Hometown. From there it crosses into Carbon County's Mahoning Township, where it flows past its namesake borough into the Lehigh River at Nesquehoning Junction. Nesquehoning made its earliest known appearance in Easton treaty council meeting minutes taken on November 10, 1756, as "Nishamemkachton, a creek about three miles beyond [Fort Allen in present-day Weissport]" (State of Pennsylvania 1851–53[7]:317). It was next mentioned on August 19, 1762, when Iroquois speakers representing themselves and the refugee Indians living along the Susquehanna River asked the Pennsylvanian authorities not to allow settlement higher up "than Nixhisaqua (or Mohony)" (ibid.[8]:748). Appearing in its modern form on the 1770 William Scull map and as Nesquchong Creek on the 1792 Reading Howell projection, Nesquehoning became the site of entrepreneur Josiah White's first attempts to build a wing dam capable of holding back enough water to float coal boats from the shallow creek's mouth downriver along the Lehigh River. Miners erected their village of Nesquehoning at its present locale in 1831. The Nesquehoning Valley corridor later became a major anthracite mining district. The many places adorned by this name in the area include Nesquehoning High School, listed in the National Register of Historic Places in 2003.

NETCONG (Morris County, New Jersey). Netcong is a part of the place name Musconetcong, resembling what Whritenour thinks was an adopted spelling of a Northern Unami variant of a Munsee word, *neetgunk*, "fertile lowland." Today, the borough of Netcong is located along the Musconetcong River just below Lake Hopatcong. Workers employed at the nearby iron mines at Stanhope first called their village South Stanhope. They adopted the contraction Netcong for the post office opened at the locale in 1889. Its status became formalized when the borough incorporated in 1894.

NETIMUS (Pike County, Pennsylvania). Lake Netimus is the focal point of Camp Lake Netimus, a Pocono-area girls' camp established in 1930 in Dingman Township several miles southwest of the borough of Milford. The camp is named for the Forks sachem Nutimus, who was active in intercultural affairs in the area during the mid-1700s. Heckewelder thought his name was Delaware for *nütamæs*, "a striker of fish with a spear."

NEVERSINK (Orange, Sullivan, and Ulster counties, New York). Today, Neversink is most notably the name of a river, a reservoir, a town, and a village in New York. The name Neversink has often been confused with similar-sounding Navesink in Monmouth County. Confusion was not lessened by the habit Monmouth settlers had of spelling their community's name in ways both similar and identical to New York's Neversink at various times (most frequently during the 1680s). The existence of identically spelled names in both New York and New Jersey has helped fuel traditions supporting the existence of an Old Mine Road stretching from the Delaware Water Gap to Kingston, New York, for more than a century.

Nearly all of the earliest occurrences of the name in New York clustered around the Catskill Mountain village of Neversink Flats, established shortly after the end of the Revolutionary War. Before then, most colonists knew the Neversink River as the Maheckkamack or Maghaghkamack Branch of the Delaware River (Reading 1915:95–106). The community known as Maghaghkamack at the mouth of the river, settled by Dutch colonists during the first decades of the eighteenth century, is today known as Port Jervis. Farther upriver, ponds built at the heads of several stretches of rapids at and around the flats at Neversink provided water that powered mills that quickly turned the area into a major logging and tanning center. The town of Neversink was established around the village in 1798, and post offices were opened at Neversink Falls in 1820 and at the village of Neversink in 1828 (Kaiser 1965). The latter locale was one of two villages flooded by the waters impounded within New York City's Neversink Reservoir shortly after construction commenced in 1941. The relocated Neversink village today stands above the dam completed in time for the reservoir's opening in 1953.

The fifty-five-mile-long Neversink River flows into and through the Neversink Reservoir along a relatively narrow valley between Ulster and Orange counties. Its uppermost East and West Branches run along the lower lip of the ancient crater left by a meteor that struck Ulster County 375 million years ago. The branches meet just across the Sullivan County line at the village of Claryville, where their waters flow past the East and West Delaware Aqueduct tunnels into the Neversink Reservoir. Those waters not pumped from the reservoir into the Neversink-Rondout Aqueduct are sluiced into the river that continues to flow south through the resort villages of Woodburne, Fallsburg, and South Fallsburg into the deep gorge protected within the New York Department of Environmental Conservation's Neversink River Unique Area. From there the river turns southeast on its way to the Orange County village of Cuddebackville, where the steep ridge of the Shawangunk Mountains forces it to make a sharp right turn south toward its junction with the Delaware River seven miles beyond at the city of Port Jervis.

NIMHAM (Putnam County, New York). Today, Nimham is the name of a cluster of recently renamed places honoring the memory of Wappinger sachem Daniel Nimham (Grumet 1992; Smith 2000) in the Putnam County town of Kent. Nimham led efforts to recover Wappinger lands in the county, an undertaking that led him to London. There officials turned the matter over to Royal Indian Superintendent Sir William Johnson, who decided against the claim in 1765. Nimham divided his time between his old home at Wiccopee in Putnam and refugee Indian settlements at Stockbridge, Massachusetts, and Indian towns at and around Oquaga along the North Branch of the Susquehanna River. He was killed along with his son Abraham (who commanded the detachment of riflemen known as the Stockbridge Indian Company) while fighting alongside other Continental Army units defeated at the Battle of Cortlandt Ridge in present-day Bronx County on August 31, 1778.

Nimham's name and memory were largely forgotten until the Civilian Conservation Corps and the state of New York erected an eighty-foot-tall fire tower atop the 1,320-foot-high hill they christened Mount Ninham in 1940. The tower and its hill, formerly known as Big Hill and Smalley Mountain, ultimately became focal points of preservation efforts that resulted in the state's acquisition of what presently is the New York State Department of Environmental Conservation's 1,023-acre Ninham Mountain State Forest and its Ninham Multiple Use Area on the mountain itself. The name also adorns several streets and roads in and around the state forest.

NIPOWIN. See **RIPPOWAM**

NISHUANE (Essex County, New Jersey). Whritenour proposes that Nishuane may be a Munsee word, *niishuahne,* "double stream." Today, Nishuane is the name of a brook, a park, a school, and several streets in Montclair Township. It appears to have been drawn from Indian deeds noting the participation of Claes Neshawan as an Indian interpreter on June 6, 1695 (New Jersey Archives, Liber E:306–307), and as the witness Nihicowen to the August 13, 1708, Mackseta Cohunge purchase (ibid., Liber I:210–11). The name apparently was another of Native culture broker Claes's names or nicknames. Nishuane's association with Claes the Indian, as well as its identification as another Munsee term for the number two mentioned in several published Delaware word lists, almost certainly eased Nishuane's way onto modern maps as the name of a tiny headwater of the Second River, whose waters rise in Montclair. Second River itself received its name during colonial times from its position as the second stream above Newark. Teaneck is another example of this practice, in this case, a Dutch name for the neck of the tenth stream that flowed through a predominantly Dutch-speaking part of the Hackensack Valley. Nishuane made its first appearance on local maps sometime around the turn of the twentieth century. Local residents gradually gave the name to the Cedar Street School, several streets, and the park built along the brook's banks sometime before 1940.

NOCKAMIXON (Bucks County, Pennsylvania). Heckewelder thought Nockamixon sounded much like a Delaware word, *nachaníxink*, "at the three houses or where the three houses are." Today, Nockamixon is the name of a township, a lake, and a state park. Local historian William Davis (1905[2]:38n.) stated that a private patent granted lands in what was then called the Durham tract that included present-day Nockamixon Township to Jeremiah Langhorne and John Chapman on September 8, 1717. Settlers first petitioning the province for a township patent for Nockamixon in 1742 finally succeeded in receiving one in 1746. Present-day Gallows Run was known as Nockamixon Creek during this time. The red-shale ridge known as the Palisades just east of the place where the Gallows Run falls into the Delaware River is still called the Nockamixon Rocks or Cliffs. Dams first planned in 1958 ultimately created Lake Nockamixon and the 5,283-acre state park opened in 1973 on its shores in Haycock and Bedminster townships. Camp Nock-a-mixon, a privately owned campground opened in 1981, and several other locales in the area also continue to bear the name.

NOMAHEGON (Union County, New Jersey). Today, Nomahegon is the name of a brook, a park, and a street in the village and town of Union. The name made its first and only appearance in colonial records as "Nohim [transcribed as Nolum in some secondary sources] Mehegum alias Wawahawany Creek or Brook" in the October 12, 1684, Indian deed to land between the Wickake [Weequahic], Pisawak [Passaic], and Raway [Rahway] rivers (New Jersey Archives, Liber A:262). The name now spelled Nomahegon found its way onto present-day Union County maps along with others retrieved from colonial records during the first half of the twentieth century.

NONOPOGE (Fairfield County, Connecticut). Nonopoge Road is one of the Indian names provided by the Fairfield Historical Society to the developer of the Lake Hills subdivision in 1952. Society members drew the name from a deposition made on May 5, 1684, by Nonopoge, his brother Winnepoge (Winnipauk), and Craucreeco (Cockinseko) affirming that their people's resistance against the English in the Pequot War of 1637 gave settlers' sovereignty over their lands (J. Davis 1885:121–22).

NORMANOCK. See **NAMANOCK**

NOROTON (Fairfield County, Connecticut). Today, Noroton Harbor, Noroton Neck, and the villages of Noroton and Noroton Heights are located in the town of Darien. The name first appeared in its present form and location during the 1880s when it was given to the Connecticut State Soldiers Home for Civil War veterans and the railroad station built to serve it. Although modern tradition asserts that it comes from Norporiton, an alleged Indian name for the local river not in local records, Trumbull (1881) showed that the name Noroton is probably an altered form of Rowayton.

NORRANS (Fairfield County, Connecticut). Norrans Ridge and Norrans Ridge Drive are located in the town of Ridgefield. The name first appeared in local records as Nawranawoos Ridge in 1712 (Sanders 2009). It is perhaps another name preserving the memory of the sachem Nowenack.

NORWALK (Fairfield County, Connecticut). Whritenour thinks Norwalk sounds like a Munsee word, *nalawahkuy, "peaceful land." Today, Norwalk is the name of a river, a harbor, a city, and much else in the Norwalk Valley. The name appeared early in colonial records, first recorded in the September 8, 1640, Indian deed to land between the Norwalke and Soakatuck (Saugatuck) rivers from their mouths "to the middle of the said rivers, from the sea a day's walk into the country" (in Robert Bolton 1881[1]:389–90). The name has seen constant service on regional maps ever since. Today, the twenty-one-mile-long Norwalk River rises from its headwaters at the junction of several streams flowing from ponds and swamps in the town of Ridgefield past numerous locales bearing its name (including five properties listed in the National Register of Historic Places between 1987 and 2000) to the place where its waters join those of Long Island Sound at Norwalk Harbor.

NYACK (Rockland County, New York). Today, Nyack is the name of several communities located close to one another on the western shore of the Tappan Zee along the lower Hudson River. The earliest reference to the name occurred in a deed to land at Neycusick signed on February 13, 1679 (Whitehead et al. 1880–1931[21]:225) on the north side of the then-contested border with New Jersey. It may be the same place identified as Navish, which was mentioned as the Indian name of Verdrida Hook in the June 23, 1682, Indian deed to land on the opposite bank of the Hudson River (Westchester County Records, Deed Book A:181–84). The latter name, from the Dutch for "hook or point tedious or troublesome to navigation" (Gehring 2012:personal communication) is today called Stony Point. The name Nyack later appeared south of Stony Point as Neejak in 1725 (Wardell 2009:71) and as Niack in 1764 (Green 1886:335). It was finally adopted as the name of the post office opened in the present-day village of Nyack in 1833. After a great deal of debate and much litigation, owners of the Nyack Turnpike Company finally received the go-ahead to begin construction of their road to Suffern in 1830. Enough of the road was completed to warrant opening of the Nyack Turnpike post office in 1834 (Kaiser 1965). New York State Route 59 presently follows much of the turnpike's old route. The hamlet of Nyack at its easternmost end formally incorporated itself as a village in 1883. Today, it is the center of a cluster of communities known as the Nyacks that also includes Central, South, Upper, and West Nyack. The name also adorns the sixty-one-acre Nyack Beach State Park managed by the New York Office of Parks, Recreation, and Historic Preservation. Although several other places around the region bear or have borne names similar to Nyack, their connection appears etymological rather than historical.

OCKANICKON (Bucks County, Pennsylvania). Heckewelder thought this name of a notable sachem came from a Delaware word for "an iron hook, pot hook." Ockanickon was an influential sachem from present-day central New Jersey. His brothers Mattano, Sehoppy (Heckewelder thought his name may have come from the Delaware words *schiwachpí*, "tired or staying [in one place]," or *schéyachbi*, "along the water's edge or seashore"), and several other relatives became leaders of Munsee and Northern Unami–speaking communities located between Brooklyn, New York, and Burlington, New Jersey. Ockanickon's deathbed oration taken down and published by a Quaker colonist (Budd 1685) made his name famous among colonial settlers in Pennsylvania. Today, his memory is mostly commemorated in the region in several street names and as the name of the Bucks County Council Boy Scout camp in Plumstead Township.

OENOKE. See **OWENOKE**

OQUAGA (Broome County, New York). Oquaga Creek is a thirteen-mile-long tributary of the West Branch of the Delaware River. Rising just above Oquaga Creek State Park in the town of Sanford, the creek flows south along the divide separating the Delaware and Susquehanna drainages to its junction with the West Branch of the Delaware River at the village of Deposit. The name has also been given to Oquaga Lake and Oquaga Lake Road just southwest of Deposit. Both Oquaga Creek and the present-day village of Deposit were identified as Cookhouse Creek and village in the 1792 Reading Howell map. A gazetteer published in 1813 identifies Cookquago, a seemingly intermediate spelling linking Cookhouse and Oquaga, as the name of today's West Branch of the Delaware River (Spafford 1813:73). The name Oquaga is most closely associated with the multiethnic Indian town located across the divide on a broad stretch of flats on the Susquehanna River's North Branch floodplain at and around the village of Windsor. Emerging as a major refugee settlement for displaced Esopus and other Munsee Indians during the 1740s, Oquaga and its nearby settlements served as a major staging area for Indian raiding parties going against the colonial frontier until a militia column burned the towns between October 8 and 10, 1778. Hostile receptions by victorious Americans forced most of the few Indians who tried to return to their lands at and around Oquaga after the war to give up and join others of their nation farther west. The group of more than one hundred refugees noted as Oquaga Indians settling at the Six Nations Reserve in Ontario during the first decade of the 1800s ultimately lost their separate identity after becoming part of the reserve community. Anguagekonk, a similar place name noted on March 12, 1702, and preserved in the form of Waughkonk Road, just northwest of Kingston, seems to come from the same Munsee etymological root as Oquaga (New York State Library, Indorsed Land Papers [3]:40). Nothing else, however, appears to link the two place names.

ORATON (Bergen and Essex counties, New Jersey, and Rockland County, New York). Today's Oraton Parkway is one of several places commemorating the Hackensack sachem Oratam, who played a prominent role in intercultural affairs in the region between 1643 and 1669. The parkway was originally designed for the Essex County Park Commission by the Olmsted Brothers firm to connect Watsessing Park in Bloomfield with Vailsburg Park in Irvington. Most of its formal landscaping was demolished during construction of the adjoining Garden State Parkway in 1955. The rerouted north and south sections of the road continue to parallel the old parkway right-of-way. Oratam also occurs as a street name in several other municipalities in and around northern New Jersey and serves as the name of a day camp operated by the Greater Bergen County YMCA in New York's Harriman State Park. Residents of Delford combined the first part of their community's name with the first syllable of Oratam to create the new borough name of Oradell in 1920. The recently sold Oritani Field Club has retained its name at the two locales in Hackensack it has occupied since 1887.

ORENECA. See **NARANEKA**

OSCAWANA (Putnam and Westchester counties, New York). Oscawana is the name of a lake and nearby places in the town of Putnam Valley and a park and nature preserve in the village of Crugers in the Westchester County town of Cortlandt. Both places are named for a prominent sachem named

Wessecano, noted in several land sales in the area between 1682 and 1690 under variant spellings of Askawanos. The name had long lain dormant in papers stored in county archives when a local entrepreneur revived it as a suitable replacement for Horton Lake where he opened his Oscawana Lake House for business during the mid-1850s. At around the same time the Hudson River Railroad opened its Oscawana-on-Hudson station across from Pegg Island several miles farther south. Pegg Island subsequently was renamed Oscawana Island. Like its namesake farther north, Oscawana-on-Hudson became a popular resort. Located in a wet town during Prohibition, Lake Oscawana drew notables such as Babe Ruth, who rented a summer cottage on the lake between 1920 and 1933. Today, Lake Oscawana is a densely populated year-round residential community. In 1958 Westchester County acquired Oscawana Island and the long-abandoned resort properties on the adjacent mainland. The town of Cortlandt currently operates the 161-acre Oscawana County Park and the nearby Oscawana Island Nature Preserve on the purchased land.

OSSINING (Westchester County, New York). Whritenour finds that Ossining sounds almost exactly like *asunung*, a Munsee word meaning "place of stones." Ossining has been fixed to a number of locales in and around Greater New York since colonial times. Today, it is most notably associated with its namesake village and town in Westchester County and the even more widely known Sing Sing Correctional Facility built in the village in 1826. Dutch chroniclers writing between 1643 and 1663 noted the existence of an Indian nation they referred to as Sintsincks on the east bank of the Hudson around present-day Ossining. The small stream that flows from the town through the village into the Hudson River has long been known as Sing Sing Brook. In 1813 the Sing Sing community became the first hamlet to formally incorporate itself as a village in Westchester. Surrounding communities split off from the town of Mount Pleasant to form the separate town of Ossining in 1845. In 1901 residents of the village of Sing Sing decided to become part of the town. Today, the name is carried on in the region by the town, the Ossining Historic District at the town center listed in the National Register of Historic Places in 1989, and a number of nearby streets and other places.

OWENOKE (Fairfield County, Connecticut). Owenoke Park is a well-preserved early-twentieth-century private beachfront community located on Owenoke Point in the town of Westport. It was named for the son of the local sachem Ponus who put his mark alongside his father's name on the July 1, 1640, deed to land in the area (in Robert Bolton 1881[2]:104). The Compo-Owenoke Historic District that includes adjacent Compo Beach was listed in the National Register of Historic Places in 1991. The name is also preserved as Oenoke, a street name that has been on local New Canaan maps for more than a century. The stretch of Connecticut State Route 124 that runs from the town of New Canaan north to the state line with New York is called Oenoke Ridge Road. The name also occurs as Oenoke Place, a small side street in the city of Stamford.

PACACK (Sussex County, New Jersey). The stream variously known as Pacack Creek and Brook flows for nearly five miles through the Pacack Valley from its headwaters in Vernon Township through Canistear Reservoir to its junction with the Pequannock River in Hardyston Township. The name first appeared in its current form at its present location in the 1778 Faden map. It may either be a close relation to Pascack farther east in Bergen County, a folk abbreviation of Pequannock (just as Pahuck may be a shortened version of Pahaquarry), an English or Scottish family name frequently spelled Pacock, Pocock, or Peacock, or something else.

PACKANACK. See **PEQUANNOCK**

PAHAQUARRY (Sussex County, New Jersey). This Indian name for the Delaware Water Gap is usually identified as a Shawnee word meaning "the place of the Pequa," a Shawnee social division. Whritenour points out that documented forms such as Pahaquoling would most closely sound like *paxkuwleew*, "it is blooming," or *paxkwuleeng*, "things are blooming," if they were Munsee words. Pahaquarry is the very recently discarded name of a township first established in 1824. Pahaquarry Township was absorbed into neighboring Hardwick Township in 1997 after its permanent population dropped to twelve residents. The name was first noted as a range of hills called Pahuckqualong in surveys performed at the Delaware Water Gap by John Reading, Jr. (1915), in 1715 and 1719. It also appeared in colonial records as the name of the place settled by the refugee Shawnee community located at the Delaware Water Gap between the early 1690s and 1727. Opinion is divided as to whether the name is a Munsee or Shawnee word and whether it was the name of the Delaware Water Gap, the Lehigh Gap, or both. The name was later affixed to the Pechoqualin Path or Lower Road, presently celebrated as an Underground Railroad route through the Poconos, and Pahaquarry Copper Mine, whose role in local Old Mine Road legends was comprehensively assessed by Kraft (1995). The old mine property was taken over by the Boy Scout council serving the greater Trenton area, who operated Camp Pahaquarra at the locale between 1937 and 1971. Although Pahuckqualong,

Pahaquarry, and Pahaquarra are no longer on present-day Delaware Water Gap maps, memories of the place live on as street names such as Pahaquarry Road in Hopatcong and Pahaquarry Street in Belvidere.

PAKANASINK (Ulster County, New York). Pakanasink Creek is a five-mile-long tributary of the Shawangunk Kill that flows north through the town of Crawford to its junction with the Shawangunk Kill near the village of Pine Bush. The name Pakanasinck first appeared in records of a Nicolls Treaty renewal meeting with Esopus Indians held on September 3, 1683 (Ulster County Records, Kingston Town Records [1]:239–40). It next appeared as Paekaensink River in New York governor Thomas Dongan's massive land purchase made on September 10, 1684 (New York State Library, Ulster County Patents:43–44). Early settlers called the Shawangunk Kill the Big Pakadasink Kill; present-day Pakanasink Creek was known as the Little Pakadasink Kill. Colonists noted that the sachem Maringomahan, who signed the 1684 deed and many others in the area between the 1680s and the 1720s, was living at Pakanasink as late as 1736. Settlers began moving to Maringomahan's old home along the modern Pakanasink Creek during the decades preceding the Revolutionary War, giving the name Peconasink to the small hamlet they built on its banks just one mile west of today's Pine Bush (Spafford 1813:297). By 1815 the people of Peconasink adopted New Prospect, the hamlet's current name, for their church and community.

PAKATAKAN (Delaware County, New York). Today, Pakatakan is the name of a 2,438-foot-high Catskill summit overlooking the villages of Margaretville and Arkville south of the East Branch of the Delaware in the town of Middletown. The uncertain location of the places identified as Pakataghkan in the July 31, 1706, Rutsen deed (Ulster County Records, Deed Book AA:400) and the evidently separate Papaconck River and Pacatachkan locales mentioned in the deed dated June 6, 1746, to land in the area (ibid., Deed Book EE:63–65), make it difficult to determine where Pakatakan began and Peapacton ended. The issue is further complicated by the fact that local residents living farther up the East Branch briefly named their first post office Papakunk in 1846 before changing it to Halcottville five years later (Kaiser 1965). This confusion was only increased by contending lawyers who pressed claims by rival landowners for lands in the region during the late 1700s. Cutting through the dross, it appears likely that today's Peapacton was the original name for the East Branch of the Delaware River and the Indian town of Papaconck on its shore. Pakatakan, by contrast, is probably located at the place identified as Pakataghkan on a 1771 survey map published in Evers

(1982) where the East Branch turns sharply north. The Pakatakan Inn subsequently built on the East Branch served tourists and travelers brought to the area first by mountain road and then, after 1871, by the Ulster and Delaware Railroad. The summer cottages of Catskill Mountain School landscape painters built around the inn after 1886 are today preserved in the Pakatakan Artists Colony Historic District listed in the National Register of Historic Places in 1989.

PAMRAPO (Bergen and Hudson counties, New Jersey). Whritenour suggests that Pamrapo may come from a Munsee word, *peemaapoxkw, "overlying rock." Today, Pamrapo is the name of an avenue at the south end of Jersey City. It also occurs as a transplant affixed to a side street fifteen miles to the north and west in the Bergen County borough of Glen Rock. The name began appearing in colonial records on the Bayonne Peninsula, first as the Pembrepock tract in a deed dated August 20, 1655 (Wardell 2009:80), and then as Pembrepogh in survey returns for lands within the tract entered between 1667 and 1669 (Winfield 1872:68–72). It continued to be used as the name of the locale straddling the border between Bayonne and Jersey City for the remainder of the colonial era. Market gardens at Pamrapo were celebrated in place names identifying the locale as Salterville (Wardell 2009:76–77), Celeryville, and finally Greenville after 1850. Initially a freestanding town, Greenville is now a neighborhood in Jersey City.

PANAWOK (Westchester County, New York). Today, Panawok is the name of a summer camp operated by the city of White Plains. The name was drawn from the listing of Indian signatories to the November 22, 1683, Indian deed to the area published in Robert Bolton (1881[2]:536).

PAPAKATING (Sussex County, New Jersey). Whritenour finds that Papakating sounds much like a Munsee word, *papakahtun, "flat mountain." Papakating presently is the name of a village and an eleven-mile-long Wallkill River tributary and its West Branch affluent that flow through the townships of Frankford and Wantage. The main stem of the creek rises in the Kittatiny Mountains just below Sunrise Mountain. Making a U-turn just east of Branchville, it passes Pellettown beyond its junction with the West Branch to the place where their combined waters fall into the Wallkill River at the present-day village of Papakating. The absence of references to names resembling Papakating in local colonial records suggests that it may be an early adaption of the place name Pahuckqualong repeatedly mentioned by John Reading, Jr. (1915), in his survey of the area in 1719. Becker (1964:55) noted that the name first appeared as "Pepper-Cotton" around the time of the Revolutionary War. It made its first appearance in its current form as the name of a small village, now called Pellettown, in Gordon's gazetteer (1834:203). A post office named Pepokating was opened at the locale by 1855. Several factories were built in the area during the nineteenth century. Better employment opportunities offered by resorts and shops in the nearby borough of Sussex (originally Deckertown) evidently drew off much of the village's population. Today, the quiet, rural, dairy-farming community of Pellettown bears only a passing resemblance to its former incarnation as bustling Papakating village.

PAPURAH (Fairfield County, Connecticut). Papurah Road is located in the Lake Hills development built in 1952 in the town of Fairfield. Members of the Fairfield Historical Society provided the developer with the somewhat respelled name that had originally appeared as Pupurah in the October 6, 1680, Indian deed and as Papuree, the lead signatory to the April 28, 1684, confirmation of an earlier sale of land in the area (in Wojciechowski 1985:101, 105).

PARAMUS (Bergen County, New Jersey). Whritenour suggests that Paramus may be a joining of the Dutch loanword *pruim, "plum," with a Munsee word, *shiipoosh*, "creek." Today, the borough of Paramus is the location of one of the nation's largest shopping-center districts. It is also famous as the place where blue laws first enacted a year after the center opened banned "worldly employment" on Sundays to keep roads otherwise congested by shoppers' cars clear for local residents at least one day a week. The name Paramus first appeared as the Pamaraqueing River mentioned in the Indian deed to land in the area dated November 11, 1709 (New Jersey Archives, Liber I:319–21), and as Paramp Seapus in a subsequent May 9, 1710, transaction (ibid.:317–19). Today, Pamaraqueing and Paramp Seapus are known as the Saddle River. The Paramus post office was opened to serve the local community in 1797. Paramus remained a quiet agricultural hamlet until burgeoning development stimulated by the growth of local market gardens thought to have inspired the Garden State's nickname led local residents to incorporate their village as a freestanding borough in 1922 (Wardell 2009:76).

PARSIPPANY (Morris County, New Jersey). Whritenour finds that Parsippany sounds much like a Munsee word, *paasihpunung*, "place of swollen tubers." Parsippany is presently the name of a village, a lake, several roads, and a number of other places in Parsippany–Troy Hills Township. John Reading, Jr. (1915:35), first recorded the name Perseapany in an April 21, 1715, entry in his survey book. Gordon (1834:203–204) noted Parsipany as the name of a creek (one of today's more southerly headwaters of Troy Brook) and as the name of a small farming and iron-making village on its shores. By midcentury, Parsippany was among the many communities along the Central Railroad of New Jersey main line that had become popular tourist resorts. Subsequent transformation of summer cottages into year-round homes around the turn of the twentieth century resulted in a population boom that led to incorporation of the expanded Parsippany–Troy Hills Township in 1928. Today, the township is a densely populated community of residences, light industrial parks, and corporate headquarters centering around the heavily traveled intersection of Interstates 80 and 287.

PASCACK (Bergen County, New Jersey, and Rockland County, New York). Whritenour thinks Pascack sounds like a shorn version of a Munsee expression, *(eenda) skapaskahk*, "where there is wet grass." Pascack Brook is an eleven-mile-long waterway that rises in New York just north and west of the village of Spring Valley in the town of Ramapo. From there it flows south into the town of Orangetown, where it passes through the ninety-seven-acre Pascack Valley Town Park. It then crosses over into New Jersey, where it flows through several localities until the Musquapsink Brook joins it just above Bergen County's 137-acre Pascack Valley County Park. From there it is little over a half mile to the place where Pascack Brook falls into the Hackensack River at the north end of the Oradell Reservoir. Early references to Pascack include mention of a Peskeckie Creek in an Indian deed dated October 16, 1684 (Whitehead et al. 1880–1931[21]:73), identification of an "Indian field called Pascaik" in a November 11, 1709, survey return (New Jersey Archives, Liber I:321–22), and notation of a river called Pasqueek in the May 9, 1710, Ramapo Indian deed (ibid.:317–19). John Campbell operated a wampum factory manufacturing shell beads distributed to Indian tribes by the U.S. government at the village called Pascack, which grew up along the banks of what locals called the Big Pascack River just below the New York border. Villagers formalized the name when they gave it to the Pascack post office, which opened in 1827 in what was then Passaic County. That year another community farther downriver in Bergen County gave the same name to its new post office. Both places coexisted fitfully even after Passaic County's Pascack became part of Bergen County when borders shifted in 1843. The figurative shoe of the upriver community dropped first when residents renamed it Park Ridge in 1870 (Park Ridge became a separate borough in 1894). Twenty years later, residents of the downstream hamlet of Pascack changed its name to Woodcliff in 1891 and, finally, to Woodcliff Lake in 1907 (Wardell 2009:77–78). Although no municipality currently bears the name Pascack, it serves as the less formal regional designation

Pascack Valley. Many commuters are familiar with this usage as the name of New Jersey Transit's Pascack Valley Railroad line. The thirty-one-mile-long route running from Hoboken to Spring Valley was first built in 1856 by the Hackensack and New York Railroad. Ownership passed through the boards of several private companies until 1976, when Erie Lackawanna Railroad executives turned the line over to Conrail, which in turn transferred its operations over to New Jersey Transit in 1993.

PASSAIC (Bergen, Essex, Hudson, Morris, Passaic, Somerset, and Union counties, New Jersey). Whritenour states that it almost certain that the meaning of the Delaware word *pasaic* or *pasáiek*, which Heckewelder first translated as "valley," is reproduced in its Munsee cognate *pahsaayeek*. Nora Thompson Dean provided the same translation for a Southern Unami word, *pahsaëk*. Today, Passaic is the name of a river, its valley, and a substantial number of places located in and around the valley's 935-square-mile drainage area in north-central New Jersey. Fast-running Passaic River feeder streams, such as the Pequannock, Pompton, Ramapo, and Rockaway rivers, flow through the valley's mountainous upper reaches. Lower tributaries of the river, such as Loantaka Brook and the Whippany River, flow into the shallow bowl-like marsh-filled depression left by the receding waters of glacial Lake Passaic at the end of the last ice age.

The ninety-one-mile-long Passaic River was first mentioned in colonial records as the Pessayack River in the July 11, 1667, Newark Indian purchase (New Jersey Archives, Liber 1:270[69]). The same document acknowledged the presence of the Pesayak Indian town somewhere along the river's lower reaches. References to places named Passaic along the variously spelled river have occurred with undiminished frequency up to the present day. One of the river's most dramatic features, the Passaic Falls (whose vertical drop rivals Niagara's), held back water that once powered the mills running the Society for Useful Manufactures factories, first chartered in 1792 at Paterson. City fathers of present-day Passaic did not give up their community's original name, Acquackanonk, until 1873. The renamed city of Passaic carried on the valley's reputation as a center for heavy industrial production well into the post–World War II era. A symbol of the decline that set in during the 1960s, the old industrial cities of the Passaic Valley are currently reinventing themselves as multicultural mixed residential/light industrial communities.

PATAUKUNK (Ulster County, New York). Whritenour thinks that the earliest recorded form of the name, Papapakapochke, sounds similar to a Munsee word, *papaa-pakaapoxkuw*, "flat rocks all around." Today, Pataukunk is a small crossroads hamlet in the town of Rochester located just south of the place where the waters of the Fantinekill fall into Mombaccus Creek to form the Mill Brook. The name's earliest mention occurred in a July 6, 1705, Indian deed as the name of a tract whose bounds extended "from Hoonckh up said creek to Nawesinck . . . at least three English miles" (Ulster County Records, Deed Book AA:353). Around a year later, another Indian deed dated November 2, 1706 (this one signed by Shawachkommin, the Indian name of the local sachem Jogee), located Papapakapochke as a place two miles above "Hoonckh" owned by a "Waggachkemeck Indian called Orekenawe" (ibid.:401). Papapakapochke was evidently one of the more northerly places where Orekenawe, already noted as the much-traveled sachem Nowenock, made his home. A small village rose along the wagon road running past a mill built at the site of Orekenawe/Nowenock's old home during the 1700s. Becoming a local center of the logging and tanning industry, the little village grew large enough to support a post office named Pataukunk that opened sometime between 1855 and 1882. The locale transformed itself into a resort destination like so many other Catskill Mountain towns around the turn of the twentieth century. Today, Pataukunk mostly serves as a residential community along Ulster County Route 3.

PATTHUNKE (Westchester County, New York). Today, Patthunke is the name of a summer camp operated by the village of Scarsdale. The name was drawn from numerous records referring to the local sachem bearing the name mentioned in deeds selling land in and around Westchester between 1666 and 1714.

PAUNACUSSING (Bucks County, Pennsylvania). Paunacussing Creek is a three-mile-long tributary of the Delaware River in Buckingham Township. MacReynolds (1976:288–89) notes that the name first appeared at its present locale in a 1703 survey as Paunaucusinck Creek. He also noted that the stream and the small community near its head were known as Milton's Creek during the early nineteenth century. Local residents subsequently resurrected the creek's Indian name around the time they renamed the hamlet Carversville for the local post master in 1833. Today, the name Paunacussing graces the creek and the 102-acre Paunacussing Creek Preserve managed by the Natural Lands Trust along its lower course.

PAUNPECK. See **PEAPACK**

PAUPACK. See **WALLENPAUPACK**

PAXINOSA (Northampton County, Pennsylvania). Paxinosa was a Shawnee leader whose memory is preserved in the names of a modern-day neighborhood, street, and school in the city of Easton and adjoining Forks Township. The name hearkens back to the time when Paxinosa first came to the Upper Delaware Valley with many Shawnees who left the Ohio Valley during the early 1690s. He had risen to the rank of sachem by the time the Shawnees left for new homes in the Susquehanna Valley in 1728. He was last noted as the Susquehanna Valley Shawnee elder who played a key role in restoring peace between his people and British colonists at treaty meetings held in Easton between 1757 and 1762.

PEAPACK (Hudson and Somerset counties, New Jersey; Orange County, New York). Peapack and its cognates Paunpeck and Peenpack are similar-sounding words possessing completely different etymologies in the Dutch and Munsee languages. Whritenour thinks Peapack sounds just like a Northern Unami word, *papeek*, "pond." He also thinks it likely that Paunpeck may come from a Munsee word, **paanupeekw*, "wide water." Ruttenber (1906a:225) focused on Peenpack's affinities to Dutch place names such as *Paan-pach*, "low, soft land or leased land." The name first occurred in the present-day New Jersey village of Peapack in the borough of Peapack-Gladstone, established in 1912. The earliest colonial records referred to this place as Pechpeck Town in the first Indian deed to land in area signed on October 29, 1701 (New Jersey Archives, Liber C:148–49). The locale was next mentioned as the Peapock Indian Town in another deed dated August 13, 1708 (ibid., Liber I:210–11), and as Pukpect on November 13, 1709 (ibid., Liber BBB:207–208). John Reading, Jr. (1915), repeatedly referred to the place as Pepeck and Papeck Old Indian Plantation in surveys made in 1715 and 1716. Proprietary deeds granted to settlers moving to Peapack following Reading's survey excused buyers from quit rents required in other parts of East Jersey. This attempt to lure colonists to the stone-strewn highland locale soon succeeded in drawing farmers to the cleared lands at the old Indian town site along the east bank of Peapack Brook just above its junction with the North Branch of the Raritan River. The village built around the locale's mill and limestone kiln gradually became large enough to support a post office of its own in 1826 (Kaiser 1965). Railroad service linked increasingly affluent Peapack residents to the docks at Hoboken and Weehawken where ferries such as the Paunpeck carried Jerseyites to and from Manhattan. The recent conferral of the name Paunpeck onto tiny Meadowlands Cromakill Creek in North Bergen almost certainly perpetuates the memory of the long-gone ferry.

Farther to the north, Huguenot refugees who had settled in mid–Hudson Valley communities such as Hurley and Marbletown moved south and west to Peenpack and other settlements along the lower course of the Maghaghkemack River (the present-day Neversink River around Port Jervis). Their little settlement of Peenpack clustered around Fort Gumaer, another of the fortifications built along the Shawangunk and Kittatiny frontier shortly after the outbreak of the last French and Indian War. The community changed its name to Cuddebackville after the Delaware and Hudson Canal was built through the area in 1828. Peenpack survives as a street name in the area.

PEENPACK. See **PEAPACK**

PEPACTON (Delaware County, New York). Whritenour has found that Papakonk, the earliest recorded form of Pepacton, sounds like a Munsee word, *peepakang*, "sweet flag (grass)." Today, Pepacton Reservoir and Dam are located along the East Branch (often called the Pepacton Branch) of the Delaware River. Also known as the Downsville Reservoir and Dam, the fifteen-mile-long Pepacton impoundment is the largest in the New York City Water System. Pepacton water destined for city faucets exits from the reservoir into the 25.5-mile-long East Delaware Aqueduct Tunnel near the place where the now-drowned village of Pepacton formerly stood. Transported to the Rondout Reservoir, water from both reservoirs flows into the 85-mile-long Delaware Aqueduct to its holding pond at the Kensico Reservoir in Westchester County. From there the water flows south into a tunnel that enters the city in the Bronx.

The Pepacton Branch was noted in Indian deeds as the Papakonk River on August 27, 1743 (Special Collections, Alexander Library, Rutgers University, Philhower Collection), and Papaconck on June 6, 1746 (Ulster County Records, Deed Book EE:63–65). Anthony Dennis noted a stream called the Popaxtunk Branch at the locale on his 1769 map. A drawing of an apple orchard marking Papekunk, just above the present Peapacton Reservoir on a 1771 William Cockburn survey map published in Evers (1982), specifically locates the home of Hendrick Hekan, a prominent Esopus leader who played a major role in land sales in and around the Catskills between 1683 and 1758. After the Revolutionary War, the locale became the logging and tanning village of Pepacton and the site of the Pepacton post office, opened sometime between 1850 and 1855. In 1942 Pepacton found itself one of the four villages located within the proposed reservoir impoundment area. Bought out by city agents, residents mostly scattered to places less likely to be threatened by future flooding. The water from the completed reservoir complex began flowing to New York City in 1955. Today, the Pepacton Reservoir supplies about one-quarter of the city's drinking water.

PENINGO (Westchester County, New York). This name occurs as Peningo Neck (also sometimes spelled Poningo), a mile-long peninsula separating Milton Harbor from the waters of Long Island Sound south of the city of Rye. Poningo is also preserved as a modern-day street name in nearby Port Chester in Westchester County. The name first appeared as Peningo, a tract on the mainland across from what was called Manussing Island, mentioned in an Indian deed to land in the area signed on June 29, 1660 (in Robert Bolton 1881[2]:130). Poningo has long been associated with Ponus, a sachem who put his mark on several land sales in the area between 1640 and 1655. A number of his kinsmen, including Taphow, signed the 1660 conveyance. The sachem's name also graces Ponus Ridge and several other nearby places in the Connecticut town of Ridgefield.

PEQUANNOCK (Fairfield County, Connecticut; Morris, Passaic, and Sussex counties, New Jersey). Heckewelder thought that Pequannock came from a Delaware word, *pekhánne,* "dark river." Whritenour demurs, suggesting instead that it sounds more like a Munsee word, **pohkawahneek,* "a creek between two hills." The name appears in a variety of different spellings in several places around the metropolitan area. In the New Jersey Skylands, Pequannock is the name of a twenty-mile-long river, a township, and much else in the Pequannock Valley. The Pequannock River rises in the Hamburg Mountains in Sussex County's Vernon and Hardyston townships. Crossing into Passaic County at Stockholm, the river forms the boundary between Passaic and Morris counties as it flows past Newfoundland, Butler, and Riverdale to its junction with the Pompton River at Pompton Plains. The stream was first mentioned in colonial records as the Poquanock River in an April 1, 1694, Indian deed to land along its lower reaches (New Jersey Archives, Liber B:651). It was next mentioned in two deeds executed in 1709, first as the Pekqueneck River on September 16 (ibid., Liber E:306–307), and next as the "Peaquaneck River also called Haysaghkin" on November 11 (ibid., Liber I:319–21). John Reading, Jr. (1915:36), identified the stream as the Poquonnonck River on April 24, 1715.

Colonists moving into the highland New Jersey village they named Poquanock near the mouth of its namesake river mostly came from Dutch- and English-speaking communities along the lower Passaic and Hackensack rivers. In 1798 the village joined together with the neighboring community of Pompton Plains to form the Morris County township of Pequnnack. Spelling of the name ultimately shifted to its modern form of Pequannock around the time Gordon (1834:213) published his gazeteer. Today, much of the Pequannock River Valley above the densely settled lower river lies within the fifteen-thousand-acre Pequannock Watershed owned and operated by the city of Newark's Watershed Conservation and Development Corporation.

Another spelling of the name, Packanack, graces a brook, a lake, a community, and a long narrow ridge in Wayne Township. The name's first appearances as Pacquanck, Packamack, and Pacquanack were entered into road returns made in 1760 and 1761 in the present-day Packanack area (Wardell 2009:75). Today, Packanack Brook is a three-mile-long stream that flows into the Pompton River just north of the place where the Pompton and Passaic rivers join together. Packanack Mountain, a four-mile-long extension of the Third Watchung Mountain, traverses the township's central north–south axis. A local developer damming the wide valley at the brook's midpoint in 1928 gave the name Packanack Lake to his pond and subdivision.

In Connecticut, the Pequonnock River is a seventeen-mile-long stream whose watershed represents the easternmost discernable extension of Munsee kinship networks during the colonial era. Nearly identical spellings of the name of the Pequannock River in New Jersey and the Pequonnock River in Connecticut represent just one line of evidence linking Indian people in both places. Others include co-occurrences of the ethnonyms Ramapo and Wappinger (the latter variously identified as Waping, Oping, and Pompton in New Jersey) in both places, also known as residences of such prominent sachems as Nowenock, Taphow, and Taparnekan. Colonists in Connecticut first noted Paugusset-speaking people identified as Panaquanike (soon regularized to Pequonnock) Indians on April 27, 1639 (Wojciechowski 1985:85). Other colonists moving into the area later in the century noticed that their Pequonnock Indian neighbors maintained close kinship connections with Munsee-speaking people living on land farther west on the border between New York and New Jersey (but also within Connecticut's coast-to-coast charter boundaries). Each of the colonies did its best to limit these relations by enacting laws requiring Indians to restrict land sales to clients authorized by provincial officials and by establishing other regulations prohibiting unauthorized visits from foreign Indians. The appearance of Pequonnock Indian men such as Nowenock, Taphow, and Taparnekan on the New Jersey–New York border and New Jersey Indians such as Chicken Warrup in Connecticut shows that these laws did not completely prevent marriages from crossing provincial lines. Such unions moved Pequonnock men from the increasingly cramped colonized confines of their narrow western Connecticut valleys to territory then unwanted by colonists in and around the Drowned Lands along the New Jersey–New York border—a location with more open space and fewer intrusions from Europeans. Those Pequonnocks who remained in Connecticut ultimately withdrew into reservations at Golden Hill and Schaghticoke, which endure to the present day. Today, the name in Connecticut occurs widely in and around the state's Pequonnock River Valley, where it appears as the Pequonnock River Railroad Bridge in Bridgeport, listed in the National Register of Historic Places in 1987.

PEQUEST (Sussex and Warren counties, New Jersey). Whritenour affirms that Pequest's resemblance to a Munsee word, **aapiikus*, "mouse," mirrors Heckewelder's identification of the similar-sounding place name Poquessing in Philadelphia as a Delaware word, *poquesink*, "the place abounds with mice or the place of mice," from a Southern Unami word, *poques*, "mouse." The Pequest River in New Jersey is a thirty-six-mile-long stream that rises at Stickles Pond in Andover Township just below the Sussex County seat at Newton. From there it flows south through the black dirt lands in Green Township before crossing the Warren County line into the township of Allamuchy. There, the Pequest enters the extensive Bear Swamp, exiting at the south end of Independence Township into another stretch of black dirt at Great Meadows. The Pequest River then passes into Liberty Township where it flows through the 4,811-acre Pequest Wildlife Management Area into White Township. Leaving the management area at Buttzville, the Pequest River flows past Hot Dog Johnny's on U.S. Route 46 into the borough of Belvidere, where its waters join those of the Delaware River.

A place noted as Pechquakock is located above a feature identified as Lacus (probably Lake Hopatcong) in the Jansson-Visscher series of maps as early as the 1650s (Campbell 1965). Robert Morden noted an area he identified as Puhacks on the New Jersey side of the Delaware just south of the Forks at present-day Easton on his map of 1688. John Reading, Jr. (1915:42), chronicled the earliest known clearly identifiable reference to a stream he repeatedly identified as the Paquaessing River at Pequest's current location during his first survey in the area during the spring of 1715. William Scull identified the same stream as Paques Creek in his 1770 map. The many swamps and drowned lands along much of its route discouraged all but miners who subsequently dug for iron in the Manunka Chunk and Upper Pohatcong Mountains that tower above the river's narrows below the Great Meadows. Railroads sparked the development of Hackettstown and Belvidere, two communities that soon grew to become county seats at the opposite ends of the river. Delaware, Lackawanna, and New York Railroad flatcars that had earlier carried iron ore to be smelted at the Pequest, Oxford, and other local furnaces began hauling steam dredges and shovels mounted on flatcars that drained the malarial Great Meadows sufficiently to draw farmers and dairymen to the locale by the 1860s and 1870s. Tourists traveling on the Lackawanna and Pennsylvania rail lines meeting at Manunka Chunk Junction helped built up the area's resort industry by the end of the century. Game fish released into the river from the Pequest Trout Hatchery and Natural Resource Educational Center operated by the New Jersey Department of Environmental Protection's Division of Fish and Wildlife since 1981 continue to draw sport anglers to the Pequest River Valley in considerable numbers during fishing season.

PERKASIE (Bucks County, Pennsylvania). Whritenour suggests that Perkasie may come from a Southern Unami cognate of Northern Unami word, *pilgussink*, "place of peaches." The borough of Perkasie is named for the East Branch of Perkiomen Creek, which flows along the community's southern border. In 1686 William Penn gave the name first recorded as Perkaming Creek three years earlier to the Manor of Perhomia he planned to erect for his children along its banks (Dunn et al. 1981–86[3]:132, 179). On February 17, 1700, Penn directed the survey of a ten-thousand-acre tract he dubbed the Manor of Perkesey. On October 25, 1701, he ordered that the manor be conveyed to his newborn son John (Buck 1888:242, 367). The land was subsequently divided up among all of Penn's children. In 1759 one parcel of the land was given to the University of Pennsylvania for its support provided that they never resell it to someone outside of the proprietary family.

Much legend has attached itself to the manor and the Indian town supposed to have been located within its bounds. A good bit of this can be traced to the Delaware leader Allumapies. (Heckewelder thought the name of this sachem, who was also known as Sasoonan, came from a Delaware word, *olumapisid*, "well tied, well bundled up.") Allumapies recalled sometime during the 1740s that he had been at Penn's first meeting with the Delaware sachems at Perkasie when he was a boy (in Myers 1970:83). Settlers had already moved to the place many called Perhaessing in the vicinity of the present-day borough when Allumapies made his statement. Myers deduced that the unnamed location of Penn's meeting with the Indian kings in the spring of 1683 was in fact what Allumapies called the Perkasie Indian town. In 1870 founders of the new village built along the Perkiomen Creek's East Branch adopted the then-unused name Perkasie for their community. It was given to the new village post office a year later and subsequently was chosen as the name for the borough incorporated by village residents in 1876. Today, the place name appears as a street, business, and organization name in and around the borough and is further preserved as the name of the South Perkasie Covered Bridge listed in the National Register of Historic Places in 1980.

PERKIOMEN (Bucks and Montgomery counties, Pennsylvania). Heckewelder thought that Perkioming, an early recorded form of the name, sounded much like Delaware words *pakihm-omeak* and *pakiomink*, "the cranberry place" and "the place where the cranberries grow," respectively. Today, Perkiomen is the name of an extensive river system within the Schuylkill drainage whose northeasternmost branches lie along the southwestern edge of the ancestral Munsee homeland. The name first appeared as Perkaming Creek in a letter to William Penn dated September 8, 1683 (Dunn et al. 1981–86[2]:483–84). Subsequently written down in a variety of spellings, it finally assumed its current form by 1850.

PERTH AMBOY (Middlesex County, New Jersey). The city of Perth Amboy is located at the mouth of the Raritan River where the Arthur Kill flows into Raritan Bay. Founded in 1684, the community was the capital of the proprietary province of East Jersey until it combined with West Jersey to form the royal province of New Jersey in 1702. Perth Amboy annually alternated with the former West Jersey capital of Burlington as the seat of the united province's government until 1776. Perth Amboy also had the distinction of twice being declared a city, first under royal charter in 1718 and again under the laws of the state of New Jersey in 1784. Territory including the surrounding township of Perth Amboy established in 1693 was absorbed by the city in 1844.

The place name Perth Amboy is a hybrid that joins Scottish and Delaware names. The name Perth honors James Drummond, Earl of Perth, the most influential of the Scottish proprietors of East Jersey. The story of Amboy, the Delaware half of the name, begins in an Indian deed dated December 26, 1652, to land at Ompoge at the location of the present-day city (New Jersey Archives, Liber I:9). Nearly two centuries later, Heckewelder (1834:376) wrote that the name came from

> Emboli. So called by the Indians who dwelt there. When they speak of this place they say "Embolink." This Indian name implies *hollow on the inside*. They say "Embolhallól," *hollow it out*. Embolhican is the name of a roundish adze, to work out bowls, canoes, wooden shovels, &c. I was formerly, for upwards of twenty years together, acquainted with a venerable and trusty Indian who had been born at that place, and who, when he died in 1780, was believed to be upwards of one hundred years old. He told me that the place, resembling something like a bowl, lying low and surrounded by higher hills, was therefore called Emboli.

Whritenour confirms that a Northern Unami word, *amboli*, "arched," means "something the ends or outer sides of which are higher than the middle of it." Today, the name is preserved by the city and its twin, South Amboy, on the other bank of the Raritan River.

PINE PLAINS (Dutchess County, New York). The village of Pine Plains was founded in 1740 as the place of residence occupied by Moravian missionaries working at the nearby Shekomeko Indian mission with Munsee and Mahican Indian people. Provincial authorities mistrusted the Moravians and their Indian converts from the outset. Living on what was then the far frontier of the northern British provinces, colonists in the area feared attack when war with France and her Indian allies again broke out in 1744. Giving in to their fears, local authorities suspecting that Moravians might be French spies saw to it that they were deported in 1746. Many of their Indian converts, whose numbers included Schebosch, a descendant of Mamanuchqua, an Esopus woman who had risen to the rank of sachem, followed them to the mission towns around Bethlehem, Pennsylvania. Resettled by local farmers, Pine Plains became the rural agricultural and residential community it remains today.

POCANTICO (Westchester County, New York). Today, Pocantico is the name of a six-mile-long river and a number of places located at various points along its route in the towns of New Castle, Ossining, and Mount Pleasant. The name first appeared as Pocanteco Creek in an Indian land purchase license issued on December 1, 1680 (O'Callaghan and Fernow 1853–87[13]:546), and as Pekcantico and Pueghanduck in the Indian deed secured on December 10, 1681 (in Robert Bolton 1881[1]:268–69). Writer Washington Irving (Crayon 1839:319) made Pocantico famous as the Sleepy Hollow featured in his popular Knickerbocker histories. The actual Slaupers Haven was a quiet body of water located across the Hudson at the lee of Stony Point where ships could safely pass the night. Stony Point was what Indians called Navish, known to the Dutch as the aforementioned Verdrida Hook, "tedious or troublesome to navigation," noted in the June 23, 1682, Indian deed to land across from Slaupers Haven on the east bank of the Hudson River (Westchester County Records, Deed Book A:181–84).

Today, the Pocantico River rises at Echo Lake in the town of New Castle. It then flows south into the town of Ossining, where it passes Briarcliff Manor to fall into Pocantico Lake, a now-decommissioned reservoir constructed in 1916. Running past the presently undeveloped 164-acre Pocantico Lake County Park, the river enters the town of Mount Pleasant, where it winds its way through the Pocantico Hills. From there it flows into present-day Sleepy Hollow at the one-thousand-acre Rockefeller Park Preserve, donated by the family to the state of New York in 1983. The stream then flows past the Sleepy Hollow Cemetery and through the Sleepy Hollow neighborhood in North Tarrytown, where its waters finally join with those of the Hudson River at the Tappan Zee.

POCHUCK (Sussex County, New Jersey, and Orange County, New York). Whritenour thinks Pochuck sounds almost exactly like a Munsee word, *poocheek*, "the inside corner or angle." Today, Pochuck is the name of the eight-mile-long creek that carries the conjoined waters of Wawayanda and Black creeks to the Walkill River at Pochuck Neck. The 503-acre Pochuck State Forest in New Jersey, whose lands include Pochuck Mountain and a section of the Appalachian Trail that passes across its northern slopes, and the recognizably respelled Lake Pochung are located at the southern end of the state forest. The name may come from *Pahuck.*, an abbreviation for Pahuckqualong (the mountains of the Delaware Water Gap) employed by a weary John Reading, Jr. (1915:41), in the entry he made in his journal at the end of a long day on May 18, 1715. If this is the case, it may be one of the few Indian place names directly traceable to an abbreviation in colonial records. Subsequent colonial knowledge of the Drowned Lands was as limited as the early information capable of shedding light on Pochuck's etymological origins. Most colonists avoided the black dirt's thick, malarial swamps where Pochuck Creek winds its way to join the Wallkill River at Pochuck Neck near the present-day village of Pine Island in Orange County. It was the kind of place people mostly traveled through on the few all-weather roads that crossed the area, such as the Pochuck to Goshen Turnpike, which opened in 1817 (Gordon 1834:18). Those spending longer periods of time in the region mostly worked the mines and quarries overlooking the Drowned Lands on and around Pochuck Mountain. This industry entered its most intensive period of development when the Sussex Railroad completed its line through the Vernon Valley in 1871. The railroad also brought the cars that carried steam-powered pumps and dredges that swiftly drained the Drowned Lands and brought back the fresh vegetables grown in the area's productive black dirt to markets across Greater New York.

POCONO (Monroe and Wayne counties, Pennsylvania). Heckewelder wrote that Pocono reminded him of the Delaware words *pockhanne*, "a stream issuing from a mountain," and *pokohanne*, "a stream running between two mountains." Writers frequently associate Pocono with such place names as Pahuckqualong and Pohopoco. All three occur within the present-day

Poconos resort region. The name Pocono appeared late, entered as Pokono Point (present-day Mount Pocono) on the 1770 William Scull map in today's Wayne County. Subsequent references mentioning the name most thickly cluster in Monroe County, where soldiers belonging to General John Sullivan's army noted it in their journals, maps, and correspondence as they marched through the area. Sent by Washington to destroy the Iroquois Indian towns in western New York during the summer of 1779, they wrote about the place they variously spelled Pocono, Poganogo, and Pokanose above present-day Stroudsburg while advancing north to their staging area at the southern approach to Iroquoia in Tioga (present-day Athens, Pennsylvania). Today, that portion of Sullivan's route still marked by the place name Pocono includes several lakes along the divide separating the Tobyhanna Creek flowing west into the Lehigh River and Pocono Creek, a stream that flows southeast into Stroudsburg, where it joins Brodhead Creek near its junction with the Delaware River at the Delaware Water Gap. Other places bearing the name include Mount Pocono, Pocono Manor (listed in the National Register of Historic Places in 1997), Pocono Plateau Lake, Pocono Point, Pocono Knob, and Pocono Peak Lake, the source of the Lehigh River just across the Monroe County line in Wayne County. The name also adorns Poxono Island (perhaps a kind of hybrid mixing Pocono with Paxinosa) on the Delaware River midway between Depew and Depue islands.

POHATCONG (Warren County, New Jersey). Pohatcong is the name of a creek, a mountain, and a township at the southern end of Warren County. Pohatcong Creek is a thirty-one-mile-long stream that rises at the northern end of Upper Pohatcong Mountain southwest of Hackettstown. It flows along the base of the northwestern slope of the mountain ridge through Independence and Mansfield into Washington townships. Passing through the latter jurisdiction's ninety-one-acre Pohatcong Creek Natural Area, the stream flows by the 875-foot-high Pohatcong Mountain in Franklin Township. It then passes through the township of Greenwich and on into Pohatcong Township, where it flows past the 127-acre Pohatcong Creek Wildlife Management Area before falling into the Delaware River just below the village of Carpentersville. The name first appeared as the Pokehatkong Brook, noted by John Reading, Jr. (1915:93), when he passed through the area on June 22, 1719. Anthony Dennis identified it as Pohatkunk in his map of 1769. Development began during the second quarter of the nineteenth century with construction of the Morris Canal and the Delaware, Lackawanna, and Western Railroad, both of which followed routes above the Pohatcong Creek's floodplain for the whole of the stream's course beyond Greenwich Township.

POHOPOCO (Carbon and Monroe counties, Pennsylvania). Heckewelder thought Pohopoco sounded similar to a Delaware word, *pockhápócka*, "two mountains butting with their ends against each other, with a stream between them." Today, Pohopoco is the name of a creek and a mountain ridge in the Lehigh Valley. Pohopoco Creek flows for twenty-eight miles north of the Aquashicola-Buckwha Creek system from Brodheadsville in the Monroe County township of Chestnut Hill west through Eldred Township. From there it passes across the county line into Carbon County's Beltzville Lake Reservoir, where it picks up streams running from Pohopoco Mountain six miles farther north. The creek then runs from Towamensing Township to its junction with the Lehigh River at Parryville just south of Weissport in Franklin Township. The earliest mention of Pohopoco identified it as Pocho Pochto Creek, along whose banks Indian raiders attacked and killed the members of a frontier family living twelve miles east of the Moravian mission town of Gnadenhuetten (present-day Lehighton) in December 1755 (State of Pennsylvania 1851–53[6]:758–59). William Scull noted the stream as "Poopoke or Heads Creek" on his 1770 map. The latter name is probably an anglicization of Haeth, the family name of the victims of the 1755 attack whose home was located on what was then known as Haeth Creek.

POMPTON (Morris and Passaic counties, New Jersey). Heckewelder thought Pompton was a Delaware word, *pihmtom*, "crooked mouthed." Whritenour disagrees, stating that Pompton sounds more like an anglicized version of a Munsee word, **pumbahtun*, "the down-sloping mountain." Today, Pompton is the name of a river, a lake, a borough, and several other places in and around what locals often call the waist of Passaic County where the Ramapo and Pequannock rivers join to form the Pompton River. In 1895 several communities situated along the narrowest part of the band girdling the county's waist joined together to form the borough of Pompton Lakes. The borough's boundaries presently take in Pompton village, the naturally formed Pompton Lake behind Pompton Falls (noted as Brocklet's Falls in a survey return made a year after the 1709 Indian purchase), and the hamlet of Pompton Junction that grew up around the place where the tracks of the New York and Greenwood Lake (later the Erie Lackawanna) and the New York, Susquehanna, and Western railroads intersected. Pompton Plains just across the Pequannock River in Morris County grew up around the Montclair and Greenwood Lake Railroad station of the same name, which was built in 1872 and listed in the National Register of Historic Places in 2008. The eight-mile-long Pompton River forms the border between Morris and Passaic counties across the whole of its route from Pompton Plains to its junction with the Passaic River at Two Bridges.

The Pompton River was first mentioned in colonial documents as Pontom in the Indian deed to land in the area signed on June 6, 1695 (New Jersey Archives, Liber E:306–307). Additional sales were made at what was called the Pamtam River on September 16, 1709 (Budke 1975a:94–96), and Pumpton on May 9, 1710 (New Jersey Archives, Liber I:317–19). John Reading, Jr. (1915), made repeated references to Pomptown during his surveys in the area between 1715 and 1719. The locale was home to the Pompton Indian community also known as the Opings, a name first mentioned on May 11, 1653, as Opingua (Grumet 1994). By the 1690s Pompton had become a mixed Indian community that included many Native people from neighboring parts of New York, other places in New Jersey, and sections of southwestern Connecticut. Most of these people left the area during the 1760s after accepting a cash settlement of one thousand Spanish pieces of eight, thus extinguishing all but their hunting and fishing rights in northern New Jersey offered by New Jersey provincial officials at the Treaty of Easton on October 7, 1758.

PONCK HOCKIE (Ulster County, New York). Ponck Hockie is the name of a neighborhood at the south end of the city of Kingston. Pokonoie Road, a street name of long standing in the hamlet of St. Remy in the town of Ulster, may be related to the name. Whritenour has noticed that the name sounds much like the pidgin Delaware expression *pungw haki*, "dust land." This may be the same kind of reference made to the way gnats swarm like dust clouds by the place name Punxsutawney in Pennsylvania. The name may also combine a Dutch word for point or hook with the Delaware language locative, *hockie*, "place."

PONINGO (Westchester County, New York). Poningo Neck and Poningo Avenue are located on the north shore of Long Island Sound between the cities of Rye and Portchester. The name first appeared as Peningo, a tract on the mainland across from what was called Manussing Island, mentioned in an Indian deed to land in the area signed on June 29, 1660 (in Robert Bolton 1881[2]:130). The name has long been associated with Ponus, a sachem who put his mark onto several land sales in the area between 1640 and 1655. A number of his kinsmen, including Taphow, signed the 1660 conveyance. The sachem's name also graces Ponus Ridge and several other nearby places in the Connecticut town of Ridgefield.

PONUNCAMO (Fairfield County, Connecticut). Ponuncamo Road is another of the Indian names provided by the Fairfield Historical Society to the developer of the Lake Hills subdivision in 1952. Ponuncamo was a prominent Indian participant in several land sales made in the area during the 1660s (in Wojciechowski 1985:95–96).

PONUS (Fairfield County, Connecticut). The name of this sachem, who sold several tracts of land along the present-day Connecticut–New York, border is preserved as Ponus Ridge and Ponus Ridge Avenue in New Canaan, as Ponus Avenue in the city of Norwalk, and in locales named Peningo and Poningo (described above) in Westchester County, New York.

POPHANDUSING (Warren County, New Jersey). Pophandusing Brook (sometimes called a creek) is a three-mile-long stream that flows through White Township into the Delaware River just south of Belvidere. The brook's location astride the stretch of Wisconsin Glaciation terminal moraine trending across the area played a major role in securing its inclusion in several geological survey reports published during the 1890s. Pophandusing's general similarity to the Pophannuck Creek mentioned in John Reading, Jr.'s (1915:40) journal entry for May 18, 1715, probably accounts for its current placement on state maps. Reading's description of Pophannuck as a considerable stream located about a mile below Penungachung Hill (present-day Manunka Chunk), however, much more closely matches the situation of the Pequest River several miles north of present-day Pophandusing Brook. Anthony Dennis noted a stream he called Popahanning just south of the river he identified by the name Pequest on his map of 1769.

POPONOMING (Monroe County, Pennsylvania). Heckewelder thought the place name he spelled Pauponaming sounded very much like a Delaware word, *pápennámenk*, "the place where we were gazing (looking at a strange object something new occurred to our sight)." The name was used until recently to identify Saylors Lake, a spring-fed glacial kettle hole in the old resort community of Saylorsburg. Local people still occasionally refer to the pond as Lake Poponoming.

PORICY (Monmouth County, New Jersey). Poricy is currently the name of a brook, a pond, a lane, a park, and a 250-acre park conservancy managed by local volunteers. The name first appeared as Porisy Run in a patent for land near Middletown entered on January 10, 1677 (Whitehead et al. 1880–1931[21]:26). Regarded as an Indian place name, it also strongly resembles Portici, the name of the palatial royal Italian villa located on the slopes of Mount Vesuvius and noted for its elaborately landscaped gardens. Colonists in several provinces imported the name Portici under a number of spellings to adorn their estates. Today, Poricy Brook in New Jersey is a three-mile-long stream that flows through Middletown Township from Red Hill to its junction with the Navesink River at Coopers Bridge. The park through which the brook flows is a popular local recreation area. The brook itself is perhaps best known among geologists and collectors interested in the 72-million-year-old fossilized shellfish, shark teeth, and other animal remains found along its banks.

PORT-AU-PECK (Monmouth County, New Jersey). Whritenour notes that all spellings of this place name sound very much like a Northern Unami word, *putpeka*, "deep still water or bay." Today, Port-au-Peck is the name of a neighborhood in the borough of Oceanport. The name is a French-sounding alteration of a Native place name mentioned in documents chronicling Indian land sales in the area variously identified as Potpocka on December 12, 1663 (O'Callaghan and Fernow 1853–87[13]:316–17), as Pootopecke on October 17, 1664 (Christoph and Christoph 1982:53), and as Pootapeck in the final title transfer documents on April 7 and June 5, 1665 (Municipal Archives of the City of New York, Gravesend Town Records, Deeds:73–74). The spelling of the name in the form of Port-au-Peck was first given to a hotel on the Shrewsbury River inlet of Pleasure Bay during the 1870s as a humorous pun transforming the prosaic Native name into something summoning up images of a fashionable French resort. Use of the name, eventually expanding to include the drawbridge crossing the inlet and the surrounding area, continued after the hotel was destroyed by a fire in 1922.

POTAKE (Passaic County, New Jersey, and Rockland County, New York). Whritenour thinks Pothat, an early orthography of Potake, preserved as a street name in the village of Sloatsburg, sounds somewhat like a Munsee word, *pahthaat*, perhaps a personal name meaning "he who hits someone by accident." Potake Pond and Brook (locally pronounced "po-tacky") straddle the state line separating Ringwood Township in New Jersey from the town of Ramapo in New York. The name first appeared as Pothat Creek in the Indian deed to land just north of Mawewieer (present-day Mahwah) signed on April 23, 1724 (Budke 1975a:111A). Pothat was subsequently identified as a tract of land sold by Nakama and his compatriots on March 7, 1738 (ibid.:114–16). Potake Pond was known as the site of Van Houtens Mills and also as Negro Pond (sometimes derogatorily noted, well into the twentieth century, as Nigger Pond) at the time when the Rockland County village of Sloatsburg built up the old mill's dam to create a reservoir at the locale (Wardell 2009:64, 69).

POUGHKEEPSIE (Dutchess County, New York). Local historian Helen Reynolds (1924), who devoted an entire monograph to the subject of Poughkeepsie's etymology, thought the name most likely meant "reed-covered lodge by the small water place." Today, Poughkeepsie is the name of Dutchess County's biggest city and its county seat. The name first appeared as a waterfall called Pooghkepesingh located in a tract on the east side of the Hudson River called Minnissingh in a deed of gift to land in the present-day city granted on May 5, 1683 (O'Callaghan and Fernow 1853–87[13]:57). Although its spellings often changed during colonial times, the name Poughkeepsie has remained at its present locale up to the present day.

POUGHQUAG (Dutchess County, New York). Trumbull (1881) suggested the translation for Poughquag of "a flaggy meadow," based on the name's similarity to a Nipmuck word, *Ap'paquaog*. Initial evidence for Poughquag's existence in Dutchess County comes from two references made in records documenting a land dispute with colonists across the province line in Fairfield, Connecticut. The first of these noted a locale called "Wombeeg [Wappinger] at a particular place called Pahsicogoweenog" on June 11, 1683 (in Wojciechowski 1985:107). The second, dated June 16, 1684, referred to a place near the "Hutson" River called Pawchequage (J. Davis 1885:122–37). Neither word closely resembled Poughquag, the name form adopted by immigrants from New England, who were probably more apt to trace out its more familiar, Nipmuck-like sounds when hearing such words. New Englanders began moving into the present-day Poughquag area on lands within the Beekman Patent, awarded on April 22, 1697. Sylvan Lake (earlier called Silver Lake) within the patent bounds was also known as Poughquag. The name was noted in local records as Pocghqeick in 1730 and Pegoquayick two years later. Quakers opened the Apoquague Preparative Meeting at what was then known as Gardners Hollow in 1771. Spafford (1813:131) noted Apoquague as a locality in Beekman. A post office was opened at the hamlet of Poughquag in 1829 (Kaiser 1965). Today, the hamlet of Poughquag and nearby places bearing the name lie at the center of the town of Beekman at the southeastern corner of Dutchess County.

POUND RIDGE (Westchester County, New York). Another English name for an Indian place, Pound Ridge is currently the name of a town, a village, a county park, and several nearby roads in a part of the Westchester uplands where Van der Donck placed an Indian locale he identified as Nanichiestawak in what he indicated was the territory of the Pachami and Siwanoy Indians on his 1656 map. Long tradition holds that this was the locale of the Indian town situated a hard day's march northwest of Greenwich where nearly all of as many as seven hundred people who had gathered "to celebrate one of their festivals" were killed by a force of Dutch and English troops under the joint command of John Underhill and Hendrick van Dyck in the winter of 1644 at the height of Kieft's War (Anonymous in Jameson 1909:282–85). No primary references identify Pound Ridge as the site of this turning point in the history of Greater New York. Although Indians may have sold land in the area as early as 1640, a place name containing the word Pound did not appear until June 11, 1701, when a locale called Pound Swamp was mentioned in an Indian deed to land at the uppermost part of the town of Rye (Westchester County Archives, Deed Book G:108). Comparison of the original with the transcript noting the name as Round Swamp published by Robert Bolton (1881[1]:699) indicates that the Westchester historian may not have gotten it down quite right. A place called Pond Pound Neck was next mentioned in what appears to be the same area on July 20, 1705 (in Robert Bolton 1881[2]:143). A reference to Old Pound Ridge in 1760 represents the earliest identifiable reference to the present-day name in the area. Incorporated as one of the original towns of Westchester following the Revolutionary War, residents argued over whether the name should consist of one word or two until the two-word alternative was official adopted in 1948. Ten years earlier, Westchester County acquired the land developed into today's 4,315-acre Ward Pound Ridge Reservation. The first part of the name honors the memory of prominent Westchester Republican politician Will Lukens Ward. The second carries on the local tradition holding that the old name represents a reference to an Indian hunting enclosure, or pound, used to trap game during communal hunting drives. The name is also preserved in the Pound Ridge Historic District listed in the National Register of Historic Places in 1985.

PREAKNESS (Passaic County, New Jersey). Preakness Mountain is a northern ridge of Second Watchung Mountain, mostly located at the northeastern corner of Passaic County. Early appearances of the name occurred in the forms Prekemis in 1735, Prakenas in 1766, and Preakness in 1771 (Wardell 2009:83–84). Today, Preakness survives as the name of the mountain, a brook, a village, and several other places in Wayne Township. The Preakness Ridge was once called Packanack Mountain. Another of its earlier names, Harteberg, Dutch for "deer mountain," led many to think that Preakness was a Delaware word having something to do with venison. Several writers have noted that it sounds a great deal like a Dutch word, *preek*, "to preach." Whatever its origins or etymology, this New Jersey place name traveled to the Pimlico Racetrack in Maryland, where the Preakness Stakes have been run annually since 1873. The name began its move to Maryland when a horse named Preakness owned by Milton Sanford, a breeder who maintained horse farms in Lexington, Kentucky, and Paterson, New Jersey, won the first race run at Pimlico in 1870 (Sahadi 2011:119–20). Three years later, the first Preakness Stakes was named for Sanford's winning horse.

PRESCOTT (Hunterdon County, New Jersey). Prescott Brook is a four-mile-long stream that flows into the South Branch of the Raritan River at Stanton Station in Clinton Township. The brook was among the streams dammed to create Round Valley Reservoir in 1960. Although Prescott is spelled like a European surname, it is more likely an anglicized rendering of Piskot, a Dutch nickname given to an Indian man whom John Reading, Jr. (1915:45), failed to engage as a guide during his surveys along what he called "the southerly branch of the Rarington River" on June 1, 1715. Whritenour affirms that Piskot is a Jersey Dutch word meaning "polecat or skunk." The name in the form of Piscot Brook most recently appeared in Gordon's (1834:217) gazetteer.

QUAKAKE (Carbon and Schuylkill counties, Pennsylvania). Heckewelder (1834:358) thought Quakake sounded like the Delaware words *cuwékeek* or *kwékêêk*, "piney lands." Quakake Creek, a name that Heckewelder reconstructed as *kuweuhanne*, is a nine-mile-long stream that flows from its headwaters in Schuylkill County through Carbon County to its junction with Hazle Creek at the village of Weatherly. Known below Weatherly as Black Creek, the creek flows four miles farther east to its confluence with the Lehigh River at Penn Haven Junction. The name first appeared as Queekeek in a 1758 report on diplomatic affairs on the Pennsylvania frontier (Hazard 1852–60[3]:413). The present-day Spring Mountain, which forms the divide between the Lehigh and Susquehanna drainages, was noted as Quakake Hill in 1787 (ibid. [6]:131). Reading Howell placed what he identified as Quacake Creek and the Quacake Valley at their present locales in his map of 1792. Colonists used the pass through a gap in Quakake Mountain along present-day Pennsylvania State Route 93 as a high road to the Wyoming Valley. Closely associated with the region's coal mining industry, the name continues to adorn Quakake Creek, the hamlet of Quakake Junction on the creek's banks in Schuylkill County, and Quakake Lake in the county of Carbon.

QUAROPPAS (Westchester County, New York). Today, Quaroppas Street and municipal day camp are located in the city of White Plains. The name was first recorded in an Indian deed dated November 22, 1683, to a tract "commonly called by the English the whit plaines, and by the Indians Quaroppas" (in Robert Bolton 1881[2]:536).

QUASSAICK (Orange and Ulster counties, New York). Quassaick (officially spelled Quassaic by the U.S. Board on Geographic Names) is the name of a small stream that rises in the town of Plattekill in Ulster County. It flows south from Plattekill into the town of Newburgh, where it forms the city of Newburgh's border with New Windsor before debouching into the Hudson River. The area drained by present-day Quassaick Creek was included within the vast tract of land taken up by Governor Dongan's Indian deed on September 10, 1684. Examination of several surviving states of the Dongan deed and its administrative recording documents failed to reveal any name resembling Quassaick. It began appearing during the first decade of the eighteenth century in documents identifying a tract containing 540 acres located at a place called "the parish at Quassaick" in present-day Newburgh, set aside for prospective Palatine German immigrants. The first Palatine settlers bypassed Quassaick, however, and instead went straight up to Dutchess County after arriving in New York in 1710. A small number subsequently settled for a time near Newburgh in an upland locale they called Quasek. Ruttenber (1906a:129) thought that Quasek and Quassaick may have been variant spellings of Cheescocks. Also known as Chambers Creek, the stream has been called Quassaick Creek since the mid-nineteenth century.

RAHWAY (Essex and Union counties, New Jersey). Whritenour thinks that the spelling of Rahway strongly resembles an anglicized representation of a Munsee word, *lxaweew*, "it is forked." Today, Rahway is the name of a twenty-four-mile-long river and the city incorporated in 1858 where it flows into the Arthur Kill marshlands. The river's two main branches rise to the west of the city in the Watchung Mountains. The most northerly of these, called the West Branch, is mostly an upland stream that flows from Verona south through the highlands of West Orange and the South Mountain Reservation into the coastal plain through a gap in the Watchungs at Millburn. The East Branch is a stream that flows from Montclair south through Orange and South Orange past Vauxhall to its junction with the West Branch at Springfield. From there the main stem of the Rahway River passes through Union, Kenilworth, and Cranford until it reaches the city of Rahway and its outlet across from Staten Island.

The earliest references to the Rahawack River and the Rawack Meadows occur in deed patents confirmed on March 18, 1670 (Whitehead et al. 1880–1931[21]:10). The present-day location of the city of Rahway went through a number of name changes during colonial times. The most celebrated of these was Spanktown, the name of the locale when British and American troops fought several sharp skirmishes in the area in 1777 during the Revolutionary War. After the war, Rahway became an industrial center and the site of an early national mint where the first coins bearing the motto "E Pluribus Unum" were struck. Development intensified rapidly following construction of rail lines across the area. One of these, the Rahway Valley Railroad built in 1897, which connected several major rail lines, operated for many years as one of the nation's most successful short lines. The Union County Park Commission's Rahway River Parkway completed in 1925 became a showpiece of American landscape architecture. Today, efforts continue to revitalize the Rahway city center and the Rahway Valley Railroad whose tracks last saw service in 1992.

RAMANESSIN (Monmouth County, New Jersey). Whritenour thinks Ramanessin means "paint stone," from the Munsee words *wallamman*, "paint," and *achsin*, "stone." Rementer suggested a Southern Unami equivalent, *olàmànahsen* (in Boyd 2005:438). In the same source Rementer thought that it also might mean "place of fish," a combination of *namèes*, "fish," and the locative suffix *ink*. Often called Hop Brook, Ramanessin Brook is a seven-mile-long tributary of the Swimming River in Holmdel Township. The brook rises at Telegraph Hill at the northeastern end of the Mount Pleasant Hills. From there it flows south through the 227-acre Ramanessin Tract Preserve and across the village of Holmdel into the Swimming River Reservoir. Sachems of a place identified as Ramanessin on June 18, 1675 (New Jersey Archives, Liber 1:290[49]–289[50]), Ramesing on May 22, 1676 (Monmouth County Records, Deed Book B:11–14), and Wromansung both on August 12, 1676 (New Jersey Archives, Liber I:401–402), and September 29, 1676 (Monmouth County Records, Deed Book B:33–35), participated in land sales around the area that ultimately culminated in the sale of their town site, identified as Romanese, on March 23, 1677 (New Jersey Archives, Liber 1:365[74]). The name has shared its position on local maps along with its other name, Hop Brook, since colonial times.

RAMAPO (Fairfield County, Connecticut; Bergen County, New Jersey; and Orange, Rockland, and Westchester counties, New York). Whritenour finds that Ramapo sounds much like a Munsee word, **alaamaapoxkw*, "under the rock." Today, Ramapo is the name of a Hudson Highlands ridge, a river, and a number of other places along the Ramapo Valley on both sides of the New York–New Jersey state line. It is also a street name in present-day Fairfield

County, Connecticut, where other New York–New Jersey frontier place names such as Pequannock and Oping originated. The name first appeared as Ramapough, one of the tracts sold by Indians living along the lower reaches of New Jersey's Ramapo River on August 10, 1700 (Budke 1975a:77–78). It was next mentioned on September 30, 1708, as the Ramapoo Indian community on the Connecticut–New York line represented by the sachem Katonah in the last deed he signed to land in the town of Ridgefield (in Robert Bolton 1881[1]: 392–93). Indians sold the bulk of their remaining lands on what was identified as the Romopuck River in New Jersey in deeds signed on November 18, 1709, and May 9, 1710 (New Jersey Archives, Liber I:317–21). Colonists retaining the name for the river immediately adopted Ramapo as the name for the community they established along its banks.

Today, the stream known as the Ramapo River is a thirty-mile-long Piedmont creek that flows from its headwaters in the New York town of Monroe south along a course that winds between the New York State Thruway and New York State Route 17 through the Ramapo Pass across the Ramapo Mountains into New Jersey. Passing Ramapo State College, the 3,313-acre Ramapo Valley County Reservation, and the 4,268-acre Ramapo Mountain State Forest, the river finally joins with the Pequannock River at Pompton Plains to form the Pompton River, whose waters finally join the Passaic River on their way to Newark Bay and the Atlantic Ocean.

Geologists define the Ramapo Mountains as a long, narrow stretch of the Hudson Highlands whose main escarpment follows the trough formed by the Ramapo Fault. Originating as an extension of foothills comprising the uppermost part of the much-younger Watchung range, the Ramapos stretch north and east across the southern border of New York's Harriman State Park to Stony Point, where they plunge under the Hudson River across from Indian Point. Ramapo also appears on present-day New York road maps as the name of a town in Rockland County that was called New Hempstead in 1791 and Hempstead in 1797 before adopting its current name in 1828 (Wardell 2009:85). In addition, the name has been given to lakes, ponds, streets, subdivisions, and other places in the area and has been adopted by the Ramapough Mountain Indian nation, whose members trace their lineage back to Indian ancestors.

RANACHQUA (Bronx and Sullivan counties, New York). Ranachqua is variously thought to have been the Indian place name for the southwest corner of present-day Bronx County purchased by Jonas Bronck in 1639 or, perhaps just as likely, as the river identified as the Aquehung or Bronxkx mentioned in the March 12, 1663, deed to West Farms abstracted by Robert Bolton (1881[2]:433–34). Today, the name appears on Bronx maps as the recently christened Ranaqua Park, a small recreation area in the Mott Haven section of the borough. Like several other Indian names in Greater New York, Ranachqua has been an internal migrant, given to a succession of Boy Scout camps operated by the Bronx Council, most notably at Lake Kanowahke (today spelled Kanawauke) in Harriman State Park from 1917 to 1928, and on Lake Nianque in the Ten Mile River Scout Camps in Sullivan County from 1929 to 1993. Today, the Hudson Valley Boy Scout Council leases Ranachqua as their primary summer camp.

RARITAN (Hunterdon, Mercer, Middlesex, Monmouth, Morris, Somerset, and Union counties, New Jersey). Whritenour thinks Raritan sounds like a Munsee word, *leelahtune*, "amid the mountains." Today, the Raritan River is New Jersey's largest river system. The river's main stem and its many tributaries drain a 1,100-square-mile area that includes more than one hundred municipalities in seven counties. The name also adorns the bay into which Raritan River waters flow as well as a number of places along the river's shores. Raritan first appeared in colonial records as the name of Indians who found themselves at war with Dutch colonists and their Indian allies in 1640. Many of their erstwhile Indian adversaries took refuge among the Raritans after the Dutch turned against all Native people living around Manhattan in 1643. More than a few embittered Indians from shattered communities in present-day Westchester remained in the Raritan country after the combatants signed a treaty formally ending the fighting in 1645.

Colonists were not the only people menacing Indians living in the Raritan Valley at this time. Extant records show that Raritans were compelled to accept the protection of the powerful Iroquoian-speaking Susquehannock nation, whose main town was located less than one hundred miles farther west in the lower Susquehanna River Valley. The Raritan Valley Indians continued to hold onto their lands astride the level overland corridor connecting New York Harbor with the Delaware River's head of navigation at the Falls at present-day Trenton even after their Susquehannock Indian protectors were driven from the region in 1675. East Jersey proprietors finally started buying up all the land they could along the main stem of the Raritan during the early 1680s. Negotiating more than one hundred land purchases, they managed to acquire nearly all Indian territory in the valley by the end of the first quarter of the eighteenth century.

Substantial numbers of places in a region stretching from Monmouth County west to the Musconetcong Mountains continue bear the name Raritan. These include the borough of Raritan in Somerset County, next to the county seat at Somerville that originally bore the borough's name; Raritan Township and the Raritan Valley Community College in Hunterdon County; and, most singularly, Natirar, a reverse spelling of Raritan that Walter and Catherine Ladd gave to the estate they completed in 1912 on the North Branch of the river astride the borders of Peapack-Gladstone, Far Hills, and Bedminster townships. Natirar was formerly well known as the convalescent facility for "deserving gentlewomen" that Catherine Ladd opened on estate grounds in 1908. Acquired by the Somerset County Park Commission in 2003, the 404-acre Natirar Park now serves as a community recreational facility.

RIPPOWAM (Fairfield County, Connecticut, and Westchester County, New York). Rippowam is the name of a seventeen-mile-long river that runs from the New York town of Pound Ridge into Connecticut, where it flows into the Long Island Sound at Stamford Harbor. It is also the name of a lake in the Westchester County town of Lewisboro as well as a street, a school, and a park in the area. Rippowam first appeared as Toquam, the land of Ponus who sold property there to a group of settlers from Weathersfield in what was then the New Haven Colony on July 1, 1640 (in Robert Bolton 1881[2]:104). The Weathersfield settlers arrived to take up their land at Rippowam during the following summer. Six months later, they changed the name of their community to Stamford (Huntington 1868:17, 67–68). Stamford colonists began calling the stream running through their town the Mill River as early as 1655 (Robert Bolton 1881[2]:105). The oldest records documenting local use of the name Rippowam as another name for Mill River date to the first decade of the 1900s. Even today, the lower eight-mile tidal stretch of the stream flowing from the sluiceway outlets of the North Stamford Reservoir Dam to Stamford Harbor continues to be called the Mill River. The name Toquam has also been recently revived as a street name in the town of New Canaan and as a school name in nearby Stamford. The name of the Greenwich town community of Nipowin and Nipowin Lane in Stamford may also be a somewhat garbled form of Nippowance, a local folk orthography for Rippowam.

ROCKAWAY (Hunterdon and Morris counties, New Jersey; Kings, Nassau, and Queens counties, New York). Places named Rockaway currently cluster in two locales in the Greater New York area. The earliest of these was mentioned "as a place called Reckouw Hacky" in the January 15, 1639, Indian deed to land on western Long Island (Gehring 1980:9). Whritenour thinks this name most closely resembles a Munsee word, *leekuwahkuy*, "sandy land." Colonists subsequently frequently referred to Native people living in this sandy area as Rockaway Indians. Today, the name of the Rockaway Indians adorns a beach, an inlet, and a neck; a neighborhood in Far Rockaway, Queens; the hamlet of East Rockaway in Nassau; several roads, streets, and avenues; and many other places along the south shore of western Long Island.

In New Jersey, the thirty-five-mile-long tributary of the Passaic River called the Rockaway River courses through the uplands of northern New Jersey's Morris County. Hamill Kenny (1976:96) thought this Rockaway came from a Munsee word, *lechauwaak*, "fork, branch." A stream flowing into the South Branch of the Raritan River in Hunterdon County called Rockaway Creek probably shares the same etymology. Observing that the Munsee word for "fork" would sound more like *leexaweek*, Whritenour thinks that the two New Jersey streams and the identically spelled places around Jamaica Bay in western Long Island both resemble what a Munsee word, **leekuwi*, "sandy or gravelly," may have sounded like to Europeans.

Colonists in New Jersey who wrote deeds transferring Indian title to lands along the lower reaches of the Rockaway River set down the name of the stream in forms such as Rachawak on November 10, 1701 (New Jersey Archives, Liber H:37–39), and Rechawak on July 29, 1702 (ibid., Liber M:555–56). John Reading, Jr. (1915:37), surveyed a substantial number of lots at various places along what he called the Rackoway and Rockaway River in the spring of 1715. Other occurrences of the name in present-day New Jersey include the borough of Rockaway, along a section of the Morris Canal in the area that paralleled the Rockaway River; Rockaway Township, farther north in Morris County; and the 2,952-acre Rockaway River Wildlife Management Area, operated by the New Jersey Department of Environmental Protection.

ROMANOCK (Fairfield County, Connecticut). Romanock Road is located in the Lake Hills development built in 1952 where Cricker Brook flows into the Samp Mortar Reservoir. It is another of the names given by the Fairfield Historical Society to the developer for his subdivision. The name comes from papers documenting a seventeenth-century land dispute in the area. On June 11, 1683, Mohegan sachem Uncas declared that his "intimate friend and acquaintance" whom he identified as Romanock lived at Sarquag (present-day Sasco). He further said that although Romanock "was a captain and of some note," he was not a sachem "that had rights to lands." Uncas ended by saying that Romanock's father lived "at or near to Wombeeg at a particular place called Pahsicogoweenog" (in Wojciechowski 1985:107). A year later, other records dated May 5 and June 16, 1684, noted that Romanock was a "foreign" war captain living in the area who also maintained a residence near the "Hutson" River at Pawchequage, at what was probably Poughquag (J. Davis 1885:122–37). Romanock was also said to have several wives, including one who had recently died at "Mawhegemuck, called Albeny." Maghagkemack, at present-day Port Jervis, was located within the bounds of what Connecticut colonists could plausibly refer to as the Albany government. Wombeeg sounds much like Wappinger, and both Pahsicogoweenog and Pawchequage resemble aforementioned Poughquag in present-day Dutchess County. These associations, should they hold up, suggest that Romanock may be a variant spelling of the name of Nowenock, the already much-mentioned sachem and namesake of such places in Greater New York as Namanock, Nanuet, and Naraneka.

ROWAYTON (Fairfield County, Connecticut). Whritenour suggests that Rowayton sounds like the Munsee words *loowathun, "it floats by," and *loowiitan, "it flows by." Today, Rowayton is a neighborhood on the north side of the city of Norwalk. The name first appeared in colonial records as "the land of Roatan," and the "creek of Rowayton called Five Mile Creek," in the Indian deed to land in the area signed on March 24, 1645 (Hurd 1881:700). Local residents resurrected the name Rowayton as a more distinctive replacement for the lackluster Five Mile Creek during the early nineteenth century. The name has since been applied to a number of local thoroughfares, the Rowayton community's millpond and dam, and a number of other places in and around the village.

ROXITICUS (Morris County, New Jersey). Whritenour thinks Roxiticus sounds something like a Munsee word, *waakwsihtukwus, "fox creek." Today, Roxiticus is the name of several streets and a golf course in the central New Jersey borough of Mendham. The first appearance of a name similar to Roxiticus occurred in an April 22, 1690, Indian deed to land at the southeasternmost corner of the present-day Nassau County, New York, town of Oyster Bay, which mentioned a river named Raskabakush running along the west bank of "Simons his Neck" (Cox 1916–40[1]:358). The Simon of the neck was John Seaman, an early settler of the Hempstead town village he named Jerusalem on the Oyster Bay border. Locals were evidently still using the name Ruskatux when talking about Seaman's Neck and its neighboring creek in 1868 when they renamed their village Seaford in honor of Seaman's hometown in the old country. Although no document thus far explicitly links Raskabakush with a Southampton-born New York colonist named Nathan Cooper, it seems that Cooper brought the name to New Jersey when he purchased a tract of land at a place identified in the deed as Roxiticus in the present-day borough of Mendham sometime between 1730 and 1740. Cooper's land was incorporated into Mendham Township upon its formation in 1749. The name of the Roxiticus village that grew up on Cooper's land was largely forgotten by the time the greater part of the township split off to form the independent borough of Mendham in 1906. Founders of the Roxiticus Golf Club resurrected the name in 1964. Roxiticus has also returned to local maps as a street name and as the informal name of the valley of the uppermost part of the North Branch of the Raritan River, which rises in Mendham.

RUMSON (Monmouth County, New Jersey). Whritenour thinks early spellings of Rumson, such as Norumpsump and Narrumson, sound much like a Northern Unami word, *nalambison*, "the belt." Today, the borough of Rumson and Rumson Road are located on Rumson Neck, a peninsula separating the waters of the Navesink and Shrewsbury river inlets. Both the inlets and the peninsula are shielded from the open waters of the Atlantic Ocean by the Sea Bright barrier beach. Rumson was first mentioned in records documenting land negotiations with Indians in the area who called the place Narowatkongh on December 12, 1663 (O'Callaghan and Fernow 1853–87[13]:316–17), and Narumsum and Narumsunk in sale papers dated April 7 and June 5, 1665 (Municipal Archives of the City of New York, Gravesend Town Records, Deeds:73–74). The locale subsequently was known by a succession of names that included Black Point, Port Washington, and Oceanic. Rumson attracted wealthy residents who began building substantial country seats along Rumson Road following the opening of rail service to the area during the mid-1800s. Local residents maintain the self-governing borough form of government adopted in 1907 to the present day.

SAMPAWAMS (Suffolk County, New York). Sampawams Point juts into the Great South Bay at the southeastern end of the town of Babylon. The name made its first appearance in colonial records as Sumpwams Neck in an Indian deed to land at the locale signed on December 2, 1670 (C. Street 1887–89[1]:171). The place was later noted on November 5, 1689, as Sampwams, the easternmost of several necks originally sold by Massapequa Indians some forty years earlier (ibid.[2]:41–43).

SANHICAN (Mercer County, New Jersey). Whritenour holds that Sanhican is almost certainly a Northern Unami word, *sanghikan*, "fire drill" (also used to refer to "fire-locks," i.e., flintlocks). Today, the name survives as Sanhican Drive in the city of Trenton. Sanhican is one of the earliest Indian place names on regional maps, making its first appearance as Sangicans at what appears to be the present-day Bayonne Peninsula on Adriaen Block's 1614 *carte figurative*. Two years later, it appeared at the Falls of the Delaware on the 1616 Hendrickszen map (both projections illustrated in Stokes 1915–28[1]:cpls. 1 and 2). Goddard's (2012:personal communication) analysis of the list of words from the Sankikans language published in 1633 by Johan de Laet (in Jameson 1909:57–60) think most belonged to what he calls the Unalachtgo dialect of Delaware, which appears to be much like Northern Unami. Heckewelder (1876:99) wrote that the Delawares sometimes referred to Mohawks as Sanhicanni. Sanhican Drive became a major thoroughfare built through a fashionable district at the west end of Trenton during the early decades of the twentieth century. It survives today as a secondary road paralleling Delaware River Drive (New Jersey State Route 29).

SANTAPOGUE (Suffolk County, New York). Santapogue Creek and its namesake streets, school, and other places are located in the Babylon town village of Lindenhurst on the south shore of Long Island. Colonists first identified the place as Santtapauge, a neck in the area included in a tract of land sold on April 19, 1669 (C. Street 1887–89[1]:134–35). Massapequa descendants of the Indians who signed the deed confirmed the sale of what they called Santapauge and two adjoining necks on July 12, 1689 (ibid.[2]:33–36). Santapogue Creek has been known by the name since colonial times.

SASAPEQUAN (Fairfield County, Connecticut). Sasapequan Road is located in the Lake Hills development built in 1952 in the town of Fairfield. Furnished at the developer's request by the Fairfield Historical Society, the name first appeared as Sasapequna, one of the Indians who signed a deed to land in the area on October 6, 1680 (Wojciechowski 1985:105).

SASCO (Fairfield County, Connecticut). Trumbull (1881) thought that Sasco came from words such as the Delaware *assiskene*, "marshy, muddy," and the Massachusett *wosoki, wosohski*, "in the marshes." Whritenour largely concurs, suggesting Sasco may be another occurrence of a Munsee word, *asiiskuw*, "mud or clay." The six-mile-long Sasco Brook flows from the town of Easton south through the town of Fairfield along the border with the town of Westport to the Buckley Pond Dam. The portion of the stream below the dam that flows as a tidal estuary into the Long Island Sound is known as Sasco Creek. Colonists repeatedly used local Indian support for Pequots defeated at a battle fought in the Sasqua Swamp in 1637 as a pretext to take their lands. The name of a creek called Sasqua was first mentioned in the March 20, 1657, Indian deed to land in the area (in Wojciechowski 1985:87). A small tract along the creek became one of the reservations that colonists set aside for use by the Sasqua Indians. Most Indians living in and around the reservation ultimately sold their lands in 1703. The majority of these people made their homes at Schaghticoke by 1736. A large part of the Schaghticoke community later joined other Christian

Indian adherents belonging to the Brothertown Movement on land set aside for them on the Oneida Reservation in upstate New York in 1785. Today, most of the descendants of these Brothertown people live in and around the Oneida and Stockbridge–Munsee Reservations in Wisconsin. Local Sasco Valley developers in Connecticut resurrected the name during the early 1900s in hopes that its romantic associations with Indians and colonial victory would attract upscale home buyers. Today, the name Sasco is preserved as a street name and as the names of the Sasco River–Kirik and Sasco Creek Marsh open spaces. Sasqua Hills, Pond, and Road in East Norwalk also maintain the name in its earlier colonial form on regional maps.

SASQUA. See **SASCO**

SAUCON (Lehigh and Northampton counties, Pennsylvania). Whritenour feels certain that Heckewelder was correct when he identified *sákunk* as a Delaware word meaning "outlet of a small stream into a larger one," and points out that its Northern Unami cognate is *sakona*. Today, Saucon is the name of a seventeen-mile-long creek and its tributaries, the Saucon Gap through which the creek flows, the townships of Upper and Lower Saucon along the waterway's banks, the hamlet of Saucon at the mouth of the creek just west of the city of Bethlehem, and a number of other nearby places. The name first appeared in nearby present-day Berks County as Sakung Creek in a July 7, 1730, declaration by Nutimus to Pennsylvania proprietary secretary James Logan that he would rather give up his lands at the locale to the proprietors than to the squatters who were then trespassing on the tract (American Philosophical Society, Logan Papers [4]:55). The proprietors subsequently purchased this land in 1732. Territory within the bounds of the 1737 Walking Purchase set up as Saucon Township in 1742 in what was then Bucks County was divided into Upper and Lower Saucon when Northampton County split away from Bucks in 1752 (Upper Saucon joined Northampton). In 1812 Upper Saucon was separated from Northampton and included in newly incorporated Lehigh County, where it remains today. The name currently also appears in slightly altered form on regional maps as Saucona Pond near Wassergass in Lower Saucon Township, where the Saucona Iron Company in the city of Bethlehem began smelting operations in 1857 (J. and L. Wright 1988:204).

SAUGATUCK (Fairfield County, Connecticut). Whritenour thinks Saugatuck sounds like a Munsee word, **nzukihtukw*, "black river." Today, Saugatuck is the name of a river, a reservoir, and a preserve west of the city of Bridgeport. The twenty-four-mile-long Saugatuck River rises at Umpawaug

Pond. From there it flows south past the Saugatuck Falls Natural Area through Weston into the Saugatuck Reservoir built in 1938. The river below the dam flows beneath two Saugatuck River bridges listed in the National Register of Historic Places in 1987 and past the Saugatuck section of the town of Westport before debouching into the Long Island Sound at Saugatuck Harbor. The name Saugatuck first appeared as Soakatuck, one of the tracts sold near Norwalk in an Indian deed dated February 26, 1640 (in Robert Bolton 1881[1]:389–90). It was identified ten years later as the Sagatuck River in another sale made on May 21, 1650 (Wojciechowski 1985:86). Like other places on or along the courses of rivers throughout the region, many locales in the Saugatuck Valley have since taken their names from the stream.

SCHAGHTICOKE (Litchfield County, Connecticut). Trumbull (1881) favored a Schaghticoke speaker's 1859 translation, *pishgachtigok*, "the confluence of two streams." The four-hundred-acre Schaghticoke Reservation in the town of Kent is the center of the present-day Schaghticoke Tribal Nation. The name also adorns locales elsewhere, most notably as identically spelled town and river names on the east side of the Hudson River above Albany, New York. The Connecticut reservation was established along the Housatonic River Valley in 1736. Indian people from western Connecticut, including many who spoke Wampano and Munsee, moved to the locale during the 1700s. Despite losses that reduced reservation boundaries to lands mostly located on rocky uplands around Schaghticoke Mountain, the people of the Schaghticoke Tribal Nation never abandoned the land and maintain it as their national center to the present day.

SCHUNNEMUNK (Orange County, New York). Schunnemunk Mountain is a six-mile-long ridge that parallels the northernmost scarp of the Hudson Highlands between Smith's Clove just outside the town of Monroe north to the parcel of public lands protected within the 2,700-plus–acre Schunnemunk Mountain State Park, managed by the New York State Department of Environmental Conservation just south of the hamlet of Salisbury Mills. A note referring to a separate land claim settlement for a tract called Skonawonck appended to a December 1, 1682, Indian deed to land farther up the Hudson Valley (Ulster County Records, Kingston Town Records [1]:117–18) may be the earliest reference to the name Schunnemunk in colonial records. The first reference mentioning a hill called Skonnemughy marking the northwest boundary of a tract including "the Murderers [now Moodna] Creeke Highland" appeared in Governor Dongan's April 15, 1685, purchase of lands immediately to the south of those acquired through his 1684 Indian deed (ibid., Patent Book 5:108–109). Mostly exploited as timberland during the past two hundred years, development has begun to intrude onto the southern and western flanks of Schunnemunk not included in the state park.

SECAUCUS (Hudson County, New Jersey). Whritenour thinks Secaucus sounds much like a Munsee word, *shkaakwus,* skunk. He also finds that it sounds similar to a Delaware word, *sekake,* "above," perhaps in reference to the high hill that towers over the meadowlands at present-day Secaucus, a freestanding town established in 1917. This peak, actually a mass of ancient volcanic rock, called Snake or Laurel Hill (after being dubbed the crowning laurel of Hudson County in 1926), is seen daily by tens of thousands of motorists cruising by Secaucus on the New Jersey Turnpike. The name first appeared in colonial records as Sikakes Island in an Indian deed to land in the area signed on January 30, 1658 (New Jersey Archives, Liber 1:3–6). Secaucus had mostly been a quiet farming and livestock-raising community for three hundred years when construction of the New Jersey Turnpike in 1952 and the relocation of the Hartz Mountain Corporation to the community a few years later sparked intense development that transformed Secaucus into what has become one of the major market centers in the Northeast.

SEEWACKAMANO (Ulster County, New York). The YMCA of Kingston and Ulster County Camp Seewackamano is located in the village of Shokan in the town of Olive. The camp is named for Sewackenamo, an influential Esopus sachem who rose to prominence as a go-between with colonists from 1659 to 1682.

SENASQUA (Westchester County, New York). Whritenour thinks Senasqua sounds similar to a Munsee word, **asunaskwal,* "stony grass." Senasqua Park is a small municipal recreational area owned and managed by the village of Croton-on-Hudson since 1960. The name made its first and only appearance in colonial records in the June 23, 1682, Indian deed to land somewhat north of the present-day park. The deed mentions a meadow the Indians called Senasqua, located on the east side of the Hudson River across from present-day Stony Point (Westchester County Records, Deed Book A:181–84).

SEPASCO (Dutchess County, New York). Sepasco is the name of a lake and a road in the town of Rhinebeck. The name first appeared as Sepeskenot in a patent to land at present-day Rhinebeck granted on April 22, 1697 (New York State Library, Indorsed Land Papers [7]:219). Palatine Germans settling at the locale around 1712 soon renamed the place Rhinebeck after a place in their homeland. The small millpond at the northeastern end of the town known as Lake Sepasco has been on local maps under that name at least since French (1860:276) published his gazeteer.

SEWANHAKA (Nassau County, New York). Following Goddard, Writenour thinks that Sewanhaka is a pidgin Delaware word, *sewan haki*, "wampum beads land." Nora Thompson Dean thought the name also sounded like a Southern Unami word, *shëwànhaking*, "at the salty place." The word first appeared as an Indian name for Long Island noted as Sewan Hacky in Indian deeds to lands in present-day Canarise dated June and July 16, 1636 (Gehring 1980:5–6), and as Suan Hacky in the January 15, 1639, Indian deed to the western part of the island (ibid.:9). Suanhacky in various forms became a popular name for ships, clubs, and businesses during the nineteenth century. The name became notorious when the steamer Sewanhaka blew up and burned, killing fifty of its passengers during an excursion cruise along the East River on June 28, 1880. Today, the name adorns a road and a yacht club in Oyster Bay and several other places in Nassau County.

SHABAKUNK (Mercer County, New Jersey). Writenour suggests that Shabakunk resembles a Munsee word, *shapakwung*, "place of mountain laurel." Today, the name is attached to two streams known as Shabakunk and Little Shabakunk creeks. Both brooks wind through the upper parts of the city of Trenton and its adjoining neighborhoods on their way to their junction with Assunpink Creek in the marshlands below East Trenton Heights. Little Shabakunk Creek is often called Five Mile Creek. The name first appeared as Shabbacunk Brook in an Indian deed to land in the area dated June 4, 1687 (New Jersey Archives, Liber M:447–49). The stream was called Shabbicunck Creek in another Indian deed signed less than a year later, on March 30, 1688 (ibid., Liber B:179–80). The name Shabakunk is perhaps best known as the stream spanned by the bridge where colonial troops successfully prevented British reinforcements from reaching the battlefield at Princeton in time to stave off defeat on January 2, 1777.

SHANDAKEN (Delaware, Greene, and Ulster counties, New York). Writenour thinks Shandaken sounds much like a Munsee word, *shundahkwung*, "place of cedar trees." Today, Shandaken is the name of several locales in and around the town of Shandaken in the Catskill Mountains. Shoheken, a name that first appeared in the area during the Revolutionary War in 1779, may be the source of modern-day Shandaken. Colonial militiamen built forts that same year at Great Shandaken (the present-day hamlet of Shandaken) and at Little Shandaken (closer to Shokan at present-day Lake Hill just west of Woodstock). Both Shandakens were part of the original town of Woodstock incorporated in 1787. Shandaken broke off from Woodstock to form a town of its own in 1804. Quickly becoming a center of the tanning industry, the hamlet of Shandaken grew large enough to support a local post office by 1815 (Kaiser 1965). Both the post office and the hamlet became known as Shandaken Centre when the village of West Shandaken, ten miles deeper into the mountains, opened a post office of its own in 1848 (ibid.). Later called Dry Brook, West Shandaken is known today as the village of Arkville in the town of Middletown.

Other places bearing the name include Shandaken Brook, which rises in the town of Shandaken and flows north into Dry Brook in the town of Hardenbergh on its way to its confluence with the East Branch of the Delaware River at Arkville. The eighteen-mile-long Shandaken Tunnel completed in 1924 transports New York City–bound water from the Schoharie Reservoir to the Esopus Creek at Allaben just east of the hamlet of Shandaken. The name also adorns the recently established 5,376-acre Shandaken Wild Forest set aside and preserved in its natural state in Catskill State Park by the New York State Department of Environmental Conservation.

SHAPNACK (Sussex County, New Jersey; Ulster County, New York; Pike County, Pennsylvania). The name Shapnack probably comes from *Schepinaikonck,* an Indian settlement north of present-day Port Jervis that appears on most editions of the Jansson-Visscher series of maps published between 1650 and 1777 (Campbell 1965). It was spelled Schackaockaninck and identified as a place at an elbow of the Neversink River in a settler's petition for lands in the Port Jervis area dated September 10, 1707 (New York State Archives, New York State Library, Indorsed Land Papers [4]:104). The name has moved around a bit since it was first recorded. Shapnack Island, sometimes spelled Shapanack, has been noted at its present-day locale on the Pennsylvania side of the Delaware River for many years. Fort Shapnack, another of the string of outposts lining the colonial frontier erected in 1756, was built across the river on the New Jersey side.

Recently, Scenic Hudson (a preservation organization that grew from the movement that stopped planned development at Storm King Mountain in 1963) imported a revised spelling of the earliest documented form of the name for their seven-hundred-acre Shaupeneak Ridge Cooperative Recreation Area located more than forty miles from the Delaware River along a small tributary of Black Brook at the northern end of Marlboro Mountain in the town of Ulster.

SHAUPENEAK. See **SHAPNACK**

SHAWANGUNK (Orange, Sullivan, and Ulster counties, New York). Local residents have pronounced the name (and sometimes spelled it) as Shongum for a very long time. Whritenour finds that the original pronunciation of the name's locative suffix *unk* produces a word that nearly identically matches the sound of a Munsee word, **shaawangung*, "in the smoky air." Today, the name is most widely associated with the Shawangunk Mountains. It is also notably linked with the thirty-five-mile-long Shawangunk Kill, which flows north and east below the mountain from its headwaters just above the New York–New Jersey state line to the village of Tuthilltown, where it joins the Wallkill River.

The name first appeared in colonial records as Sawankonck in the January 24, 1682, Indian deed to a tract of land along the lower reach of the Shawangunk Kill today known as Bryunswick (Ulster County Records, Kingston Court Records [6]:20). The earliest reference yet found by Fried (2005:6) associating the name with the ridge is the December 17, 1743, provincial assembly act declaring "the foot of Shawangough Mountain" to be the western boundary of the newly established Shawangough Precinct. This precinct became a town in 1788. A post office subsequently was opened at the hamlet of Shawangunk within the town in 1792. Local residents changed the hamlet's name to Reeveton in 1853 (Kaiser 1965). The name was changed again, this time to Wallkill, shortly after the Wallkill Valley Railroad completed the section of its line passing through the village in 1871. Today, the name Shawangunk is most widely associated with the cliffs along the highest part of the mountain ridge within Minnewaska State Park and the Mohonk Preserve.

SHAWNEE (Morris and Sussex counties, New Jersey; Monroe County, Pennsylvania). Shawnee is an Eastern Algonquian word for "southerners." The name in the Greater New York area is most commonly associated with the village of Shawnee-on-Delaware on the Pennsylvania side of the river just above the Delaware Water Gap. Shawnees mostly belonging to the Pequa division of their nation first came to the water gap between 1692 and 1694 as refugees trying to escape Iroquois attacks on their Ohio Valley communities during the first French and Indian War. Led by Paxinosa and other sachems, these Shawnees lived uneasily with their colonial neighbors at the locale until threats of attack made by Iroquois warriors and run-ins with hostile colonial neighbors compelled them to abandon their homes in the area during the summer of 1728. Most followed Paxinosa to Wyoming in the Upper Susquehanna Valley. Others moved farther west to Allegheny

Valley Indian refugee settlements in western Pennsylvania. Colonists moving onto their lands continued to use variations on the Shawnee name Pechoquealin and Pahaquarry when talking about the area. They also referred to present-day Depue Island as the Great Shawana Island. The village of Shawnee grew up on the Pennsylvania side of the Delaware River alongside what are now called the Depue and Shawnee islands. Prominent Shawnee village resident C. C. Worthington (builder of the Shawnee Inn, Shawnee Playhouse, and namesake of present-day Worthington State Park across the river in New Jersey) led the effort to give the name Shawnee-on-Delaware to the community during the 1890s to avoid confusion with other places in the general vicinity, such as New Jersey's Lake Shawnee in Jefferson Township in Morris County, and Shawanni Lake next door in Sussex County.

SHEHAWKEN (Wayne County, Pennsylvania). Heckewelder thought Shehawken came from a Delaware word, *schohacan,* "glue." The five-mile-long Shehawken Creek is a stream that flows into the West Branch of the Delaware River at Point Mountain just above the place where the Delaware's West and East branches diverge at Hancock. The name was first mentioned in colonial records in a June 3, 1751, deed to the territory between the Delaware River branches as "Shokakeen where Papagonck falls into Fishkill" (Gould 1856:242). It made its next appearance as Shouhauken in the New England purchase of land in the region made on May 6, 1755 (Boyd and Taylor 1930–71[1]:260–71). Residents referred to the hamlet on the New York side of the Delaware River forks as Chehocton (noted as Shehawkin Fork on the 1769 Dennis map) until the postmaster of the post office opened at the locale in 1815 opted for the name Hancock. The many small lakes built along the Shehawken Creek's upper reaches, whose number includes Lake Shehawken, have been resort destinations for more than a century.

SHEKOMEKO (Columbia and Dutchess counties, New York). Shekomeko is a Mahican name for a mixed Mahican-Munsee community. Today, Shekomeko Creek is a ten-mile-long tributary of the Roeliff Jansen Kill. The creek was named for the Moravian Indian mission that stood on its banks in the present-day hamlet of Bethel between 1740 and 1746. A number of the Indian converts who moved to the Shekomeko mission traced descent to Esopus and Wappinger forebears. Most joined the Moravian missionaries, who had been living two miles away in Pine Plains, in new homes in Pennsylvania's Lehigh Valley after Dutchess County authorities expelled the missionaries as suspected French spies in 1746 during the third French and Indian War.

SHENOROCK (Westchester County, New York). The hamlet of Shenorock is currently located on the banks of Lake Shenorock at the north end of the Amawalk Reservoir in the town of Somers. In 1930 developers established the Lake Shenorock Corporation, which formed the basis of the present-day community. They took the name Shenorock from the pages of Robert Bolton's history (1881[2]:150–52), which noted a sachem identified by that name in three Indian deeds to land in the area signed between November 8, 1661, and January 12, 1662. Shenorock was only one of the many spellings (others include Shanorocket and Shanorockwell) that colonists used to identify Sauwenaroque, a prominent local Indian leader who took part in land sales in and around today's Westchester County between 1636 and 1666.

SHIPPAN (Fairfield County, Connecticut). Shippan Point is a peninsula that juts out into Long Island Sound at the south end of the city of Stamford. The first mention of Shippan occurred in the July 1, 1640, Indian deed to land in the area (in Robert Bolton 1881[2]:104). Townsfolk shared land at Shippan as commonage before dividing it into private lots at the beginning of the eighteenth century. Shippan Point remained a small farming and residential community until the late nineteenth century when the opening of Shippan House and several amusement parks converted it into a popular resort. Local homeowners established the Shippan Improvement Association in 1902 to maintain Shippan Point as a quiet residential neighborhood.

SHOHOLA (Pike County, Pennsylvania; also in St. Louis County, Minnesota). Heckewelder (1834:359) thought the place name he spelled as Shahola sounded like *schauwihilla*, Delaware for "weak, faint, or depressed." Inclusion of the root word of this name in those of Delaware leaders such as Weequehela and Pachgantschillas suggests a possible allusion to the way the cares of responsibility wearied leaders. Today, Shohola is most notably known as the name of a township and the thirty-mile-long creek that flows through it into the Delaware River at the village of Shohola. The earliest known mention of the name occurred in the form of a stream noted as Sheoke just opposite another creek identified as Halfway Brook on the New York side of the Delaware River above Pond Eddy on a map drawn by Anthony Dennis in 1769. Shoholy Creek and Shoholy House, an inn on the creek about where Shohola Falls is located today, appeared one year later on the 1770 William Scull map. Local tradition holds that Thomas Quick, father of Tom Quick, the notorious "Indian Killer" celebrated until recently in entertaining stories presenting cold-blooded murders as crafty vengeance, was the first colonist to settle along the Shohola Creek. The Barryville-Shohola area became a popular rest stop for log raftsmen guiding their cargoes down the Delaware River during the first decades of the nineteenth century. Population in the vicinity began to grow after the Delaware and Hudson Canal was built through Barryville in 1829. In 1841 the Sylvania Association, a commune whose membership included New York publisher Horace Greeley, established their colony by Sylvania Lake along the Shohola Creek in the hamlet of Greeley. A bad summer that ruined commune crops put an end to the experimental community in 1845.

More permanent development began in 1848 when the New York and Erie Railroad poured more than a million dollars into major roadbed and viaduct construction along the Pennsylvania shore of the Delaware around the mouth of Shohola Creek. The growing numbers of workers and others brought in by the railroad led to the incorporation of Shohola Township in 1851. Shohola subsequently came to widespread public attention as the site of several train wrecks, the worst of which occurred on July 15, 1864, when the boiler of the engine pulling a train filled with Confederate prisoners bound for the prison camp at Elmira, New York, blew up. More than two hundred passengers were injured by the blast; fifty Confederate soldiers and seventeen Union guards were killed. The railroad continued to dominate the economic life of the area in spite of this and other disasters. In 1903 Erie Railroad passenger cars brought campers to Camp Shohola, one of the first private summer camps opened in the region. Flatcars in returning trains carried away great slabs of Shohola bluestone, which was much in demand as a durable paving stone for city sidewalks. In recent years, Shohola Township has become a mixed residential and resort community. Residents helped list the Shohola Glen Hotel in the National Register of Historic Places in 1997.

SHOKAN (Ulster County, New York). Shokan is a truncated form of Ashokan, which Whritenour thinks sounds like a Munsee word, *aashookaan*, "people are walking in the water." Today, Shokan and West Shokan are hamlets on opposite sides of the Ashokan Reservoir in the town of Olive. Both communities were among those forced to relocate to higher ground by the construction of the Ashokan Reservoir between 1907 and 1917. The Continental Congress's authorization for construction of a fort at Shoheken in 1779 represents the earliest known appearance of the name that has since been adapted and adopted at different locales as Ashokan, Shandaken, and Shokan. Residents of Ashocan changed their community's name to Caseville in 1832 (Gordon 1836:740) before finally deciding on the name Shokan in 1842 (Kaiser 1965).

SIACUS (Fairfield County, Connecticut). Siacus Place is located in the Lake Hills development built in 1952 in the town of Fairfield. Members of the Fairfield Historical Society provided the developer with the name, which originally appeared among the Indian signatories to the April 28, 1684, confirmation of an earlier sale of land in the area (in Wojciechowski 1985:101).

SICOMAC (Bergen County, New Jersey). Whritenour proposes that Sicomac may be a Munsee word, *nzukameekw*, "black fish." Today, Sicomac is a hamlet in Wyckoff Township. The name was first mentioned in reference to a tract of land called Schichamack not included in the sale of a large tract "near Pamtam" made on September 16, 1709 (Budke 1975a:94–96). Various spellings of the name Sicomac have appeared on regional maps since colonial times. It presently adorns the hamlet of Sicomac and occurs as a street, school, and business name in the area.

SINGAC (Passaic County, New Jersey). Whritenour thinks that Singac sounds like a Munsee word, *siingeek*, "outside corner or angle." Today, Singac is the name of a three-mile-long brook and the hamlet located across from the place where it flows into the Passaic River. The name first appeared as "Spring Brook or Singanck" in the June 10, 1696, confirmation to an earlier land sale in the area (Whitehead et al. 1880–1931[21]:247). It was again mentioned as Singkeek Creek in an Indian deed to a nearby tract dated September 3, 1714 (Budke 1975a:109–11). The stream today called Singac Brook is formed by the junction of the Preakness and Naachpunkt brooks in the hamlet of Preakness. The southward-flowing brook that carries the waters of these united brooks forms the border between the borough of Totowa and Wayne Township from its place of origin to the locale where it falls into the Passaic River. The hamlet of Singac at the stream's mouth is a former farming community that is now a residential neighborhood at the western end of Little Falls Township.

SING SING. See **OSSINING**

SIWANOY (Fairfield County, Connecticut; Bronx and Westchester counties, New York). An Eastern Algonquian Indian name variously translated as "southerner" or as *sewan*, another name for wampum. The name first appeared as Sywanois fixed onto a location in present-day southeastern Massachusetts noted in Adriaen Block's 1614 map. Slightly rewording Block's Sywanois in 1625, Johan de Laet (in Jameson 1909:44) wrote that Indian people calling themselves Siwanois lived along the north shore of Long Island Sound "for eight leagues, to the neighborhood of Hellegat" [modern Hellgate]. Ruttenber (1872:81–82) first identified the Siwanoys as a chieftaincy stretching along the north shore of Long Island Sound from New Haven to Manhattan across the East River from present-day Hell Gate. He also thought that they were a major component of a Wappinger confederacy stretching between the Connecticut and Hudson river valleys. Although both identifications are still very widely accepted, most specialists agree that no extant evidence identifies a Siwanoy Indian or community by name during the colonial era. Today, Siwanoy appears on local maps as a street and school name in several places on the New York–Connecticut border, as the name of a country club in Bronxville, and as the 1.8-mile-long Siwanoy Trail in Pelham Bay Park in the Bronx.

SOUTH AMBOY. See **PERTH AMBOY**

SQUAN. See **MANNASQUAN**

SQUANKUM (Monmouth County, New Jersey). Today, Squankum is the name of a small stream and two hamlets called Squankum and Lower Squankum in Howell Township. Squankum Brook joins the Manasquan River just east of Lower Squankum in Allaire State Park. The word was first mentioned in colonial records as the name of a piece of land "commonly called Squancum" in the deed to land between Crosswicks and Assunpink creeks signed on June 24, 1689, and later referred to as Squamcunck in a deed confirmation made on June 7, 1701 (Whitehead et al. 1880–1931[21]:326). A resident of Squankum named Israel Williams brought the name south with him when he moved to Gloucester County, New Jersey, in 1772. In 1842 residents in Gloucester renamed the community Williamson in his honor to avoid confusion with the Monmouth locale farther north.

STISSING (Dutchess County, New York). The borders of the towns of Milan, Stanford, and Pine Plains meet at Stissing Mountain (elevation 1,403 feet). The name also adorns Stissing Lake and other nearby locales. Stissing appeared in its current form in a surveyor's return entered in 1743 (Huntting 1897:25). The name Stissing is most notably preserved in the 507-acre Thompson Pond and Stissing Mountain Preserve managed by the Nature Conservancy and by the name of the 595-acre Stissing Mountain Multiple Use Area operated by the New York Department of Environmental Conservation. Local legend holds that the name originally belonged to a local tribe or chief named Tishasink, Teesink, or Stishink. Teesink Crossroads, a Pine Plains conservation group, adopted the name to symbolize its commitment to preserving the area's natural and cultural history.

SUCCASUNNA (Morris County, New Jersey). Whritenour thinks Succasunna sounds much like a Munsee word, *nzukasunung*, "place of iron." Today, the consolidated village of Succasunna-Kenvil is located in Roxbury Township. The name first appeared as "a brook called Sacconothainge" in an Indian deed to land in the area dated December 3, 1701 (New Jersey Archives, Liber O:145–48). The area was evidently already known as an iron-rich locale when John Reading, Jr. (1915:92), picked up samples of ironstone at the place he called Sukkasuning on May 9, 1716. Miners began digging open pits in the area soon after Reading completed his surveys. Farmers were also drawn to the unusually broad and stone-free if somewhat mucky flats at what came to be known as the Succasunna Plains. A substantial leather tanning industry emerged in the area by the dawn of the early nineteenth century. The need to transport ore, produce, and hides to market encouraged entrepreneurs to build a "turnpike from Suckasunny to Dover" in 1813 (Gordon 1834:16).

The opening of main line railroads through the area during the 1850s quickened the pace of development. Demand for explosives to break up ore-bearing veins in the ever-deepening pits stimulated construction of dynamite factories in Kenvil that were soon followed by a succession of rather devastating industrial accidents. The accident-prone explosives industry outlasted the local iron mines, whose operations finally ended around the turn of the twentieth century. Installation of the world's first electronic telephone switching system at Succasunna in 1965 has become a point of local pride. At present, Succasunna-Kenvil is a mixed residential and light industrial community.

SUCCESS (Nassau County, New York). Lake Success is a glacial kettle hole located in the village of the same name (incorporated 1926) in the town of North Hempstead. The name may well be a developer's conflation of the attractive word "success" with an appealingly romantic Indian association provided by Suscoe's Wigwam, a place in nearby Matinecock Neck long regarded as the home of the prominent Matinecock sachem Suscaneman (Grumet 1996). The somewhat more distant Ciscascata is the only other Indian place name resembling Success in colonial records. The shores of Lake Success briefly served as the temporary headquarters of the United Nations between 1946 and 1951. Today, Lake Success is mainly a community of residential neighborhoods and corporate campuses.

SYMPAUG (Fairfield County, Connecticut). Sympaug Brook is the name of a three-mile-long stream that rises in Sympaug Pond at the upper end of the town of Redding. The stream flows from Redding north through the town of Bethel to its junction with the Still River in the city of Danbury. The

Simpaug Turnpike is a public road that runs between the village center of Bethel and U.S. Route 7 in Redding along much the same route laid out by the Simpaug Turnpike Company in 1832. The origin of the name is obscure; Sanders (2009) notes that it appears as Syenpauge and as Semi-Pog, a brook in Danbury recorded in 1795. Whritenour thinks that Simpaug sounds much like Simpeck, a long-forgotten Indian place name for a mountain in northern New Jersey's Pequannock Township. He further thinks Simpeck sounds like a Northern Unami word, *msimpeekw, "hickory nut pond."

SYOSSET (Nassau County, New York). Whritenour suggests that the earliest recorded form of this name, Ciscascata, sounds like a Munsee word, *asiiskuwaskat, "it is muddy grass." Tooker (1911:255–56) thought Syosset represented the way its first recorded name, Schouts Bay (Dutch for Sheriff's Bay and now called Manhasset Bay) sounded to local Indian ears. Today, Syosset is a residential community in the town of Oyster Bay. The similar-sounding word Ciscascata first appeared in an Indian deed to land in the area signed on May 20, 1648 (Cox 1916–40[1]:625–26). Colonists called the place East Woods. Postal authorities briefly revived a respelled version of Ciscascata in the form of Syosset as the name for the post office that operated at Oyster Bay from 1846 to 1848. Nearby Glen Cove took up the name Syosset for one week before changing it back (French 1860:550). The name finally stuck in 1854 when the Long Island Railroad opened its Syosset Station at what is now the village center.

TACKAMACK (Rockland County, New York). Whritenour thinks Tackamack sounds like a Munsee word, *ptukameekw, "round fish." Both the south and north sections of Tackamack Park are managed by the Town of Orangetown Parks Department. The name is thought to be a much-altered variation of Raikghawaik, a different Munsee word that Whritenour thinks resembles *leexaweek, "that which forks." The latter was first mentioned in 1709 as a tributary of the Saddle River now called Hohokus Brook that some settlers expanded as a term identifying the entire Saddle River system (Wardell 2009:93). Tackamack is a modern orthography given to the park, established on former estate land atop Clausland Mountain acquired by the town of Orangetown by 2003.

TACKAPAUSHA (Nassau County, New York). Tackapausha Preserve and Museum are located on an eighty-four-acre tract of land along the border of the villages of Seaford and Massapequa. The property was named for Tackapausha, the sachem of Massapequa who served his people as the most prominent intermediary with colonists on western Long Island from 1643 to 1699.

The county acquired the property in 1938. Tackapausha Museum opened in 1964. Today, both the preserve and the museum are managed by the Nassau County Department of Parks, Recreation, and Museums.

TACKORA. See **NARANEKA**

TACONIC (Columbia, Dutchess, Putnam, and Westchester counties, New York). Whritenour thinks Taconic sounds like the Munsee words **takwahneek*, "adjoining stream," or **wtakwahneek*, "gentle stream." Today, the name is most closely associated with the 104-mile-long Taconic State Parkway built between 1929 and 1961 whose right-of-way extends north from the Bronx River Parkway at Kensico to its junction with the Berkshire extension of the New York State Thruway (Interstate 90) in Columbia County. Various spellings of the name presently adorn such places as Lake Taghkanic State Park; the multiple-unit Taconic State Park, one of whose units, the Rudd Pond Area in Millbrook, is within the traditional Munsee homeland; and the 909-acre Taconic-Hereford Multiple Use Area in Pleasant Valley managed by the New York State Department of Environmental Conservation. The name first appeared as Tachkanick, the location of a mixed Esopus-Mahican community whose leaders sold the land purchased by Robert Livingston as the nucleus for his Livingston Manor on August 10, 1685 (Brooklyn Historical Society, Livingston Family Papers, Folio 11).

TAKANASSEE (Monmouth County, New Jersey). Takanassee Lake is a tidal inlet located in the city of Long Branch. The name first appeared in colonial records as a place called Takanesse mentioned in an Indian deed to land in the area dated November 16, 1674 (New Jersey Archives, Liber 1:265[74]). It was next mentioned as Tanganawamese Field noted in a nearby tract sold on June 18, 1675 (ibid.:290[49]–289[50]). The name was revived in 1906 by a Jersey shore entrepreneur who built his Takanassee Hotel as a summer resort catering to visitors brought to Long Branch by the railroad. The hotel burned down during the 1930s. A namesake hotel that opened in 1921 at the Catskills resort town of Fleischmanns suffered the same fate in 1971. The Takanassee Beach Club currently occupies the site of the Takanassee Coast Guard Station (formerly U.S. Life Saving Station Number 5), which was deactivated in 1928.

TAMAQUES (Union County, New Jersey, and Monroe County, Pennsylvania). Nora Thompson Dean thought Tamaque sounded much like a Southern Unami word, *tëmakwe*, "beaver." Today, the 106-acre Tamaques Park and Tamaques Pond are located on land managed by the Westfield Recreation

Commission acquired during the early 1960s by Westfield Township. The name first appeared in the Westfield area as the Indian name of the place "called by the English the Great Swamp" in an Indian deed to land in the area dated September 14, 1677 (New Jersey Archives, Liber 1:251[88]–250[89]). The name also occurred locally as the name of an Indian man variously identified as Tamack and Tamage in deeds signed between 1668 and 1677 (ibid., Liber 1:42–43, 121–22; Liber A:328). Other places named Tamaque in and beyond Greater New York (examples include Tamaque Lake in Tobyhanna Township, Pennsylvania, and the borough of Tamaqua in Pennsylvania's Schuylkill County) evidently come from words taken from Delaware dictionaries fixed onto artificial features such as lakes and subdivisions.

TAMMANY (Warren County, New Jersey; Pike and Schuylkill counties, Pennsylvania). Tammany was a Delaware sachem who signed over several tracts of land to William Penn during the 1680s and 1690s. Heckewelder thought that a much-used spelling of his name, Tamenend, sounded very much like a Northern Unami name meaning "the affable." The sachem's name became famous as New York City's legendarily corrupt Tammany Hall political club. Tammany Hall became a byword for machine politics and graft from its founding in 1786 to its final dissolution during the 1960s. New York's Tammany Hall was only one of the many Tammany Society chapters that sprang up across the British colonies following the founding of the first Sons of King Tammany club in Philadelphia in 1772. Festivals honoring the symbol of the new American nation personified in the figure of Saint Tammany soon began to replace traditional May Day celebrations throughout the country. Tammany also became a fixture in American popular culture, lionized in songs, statues, and stage plays, and as a noble character in such novels as *Last of the Mohicans*. Places given this widely known sachem's name in Greater New York include 1,545-foot-high Mount Tammany on the New Jersey side of the Delaware Water Gap and the Pennsylvania hamlet of Tamiment across the river in Lehman Township.

TAPORNECK (Fairfield County, Connecticut). Taporneck Court is a street name in a Ridgefield subdivision built in 1982. Taporneck was the name of a prominent Pequonnock Paugusset sachem first mentioned in land sales in Connecticut in 1680. Between 1696 and 1723 he signed a number of deeds to lands along the borders of Connecticut, New York, and New Jersey under such spellings as Taparnekan, Taparonick, and Taporanecam. The spelling of his name currently on the map in Ridgefield was recently retrieved from local records and adopted at the suggestion of the local town historian.

TAPPAN (Bergen County, New Jersey; Rockland and Westchester counties, New York). Heckewelder thought Tappan sounded like a Delaware word, *thuphāne*, "cold stream issuing from springs." Whritenour thinks the name more closely resembles a Munsee word, **tupahan*, "rolling stream." Today, the name is perhaps most widely recognized as the namesake of the Tappan Zee Bridge that carries the New York State Thruway (Interstates 87 and 287) over the Tappan Zee, a mile-wide stretch of the Hudson River between Nyack and Tarrytown. Local residents living on either side of the state line separating Bergen County, New Jersey, from Rockland County, New York, also know Tappan as the name of a large local reservoir and as the name of the hamlets of Tappan, Old Tappan, and several other places in the area. The name first appeared on a navigational chart drawn in 1616. Continually on local maps since that time, Tappan should not be confused with other places sharing the same or slightly differently spelled versions of the name in and around the metropolitan area. Although a few of these latter occurrences may be imports marking some past association with the Lower Hudson Valley, most refer to colonists bearing family names such as Tappan, Tappin, and Tappen.

TAQUOSHE (Fairfield County, Connecticut). Taquoshe Place is located in the Lake Hills development built in 1952 in the town of Fairfield. Furnished at the developer's request by the Fairfield Historical Society, Taquoshe's name first appeared as an Indian signatory to the December 29, 1686, deed to land at Umpawage (Wojciechowski 1985:110).

TASHUA (Fairfield County, Connecticut). Tashua is a popular place name in and around the present-day town of Trumbull. The name first appeared in a family deed executed on May 5, 1710, which mentioned Taw-Tashua Hill. The 615-foot-high prominence now called Tashua Hill was an early colonial community focal point. Today, the name Tashua adorns a town neighborhood, several streets, two parks, a golf course, and several other places near Tashua Hill in and around Trumbull.

TATAMY (Northampton County, Pennsylvania). Tatamy is the name of a borough incorporated in 1893 in Forks Township. The borough lies just south of the three-hundred-acre Stockertown tract purchased in 1741 by Moses Tunda Tatamy, the only Indian permitted to privately acquire land within the limits of the 1737 Walking Purchase. Tatamy was a noted frontier diplomat and a prominent Christian convert from central New Jersey (W. Hunter 1996). A pass through the Kittatiny Ridge about ten miles above the borough of Tatamy is still known as Tott Gap. A consortium of preservation groups and government agencies presently manage the two-thousand-acre Minsi Lake/Tott's Gap Corridor connecting both locales.

TATETUCK (Fairfield County, Connecticut). Tatetuck Brook is a three-mile-long stream that flows into the northwestern end of the Easton Reservoir in the town of Easton. Tatetuck appeared as Tatecock Brook in 1784. It began appearing under its present spelling in geological reports published during the 1890s. Tatetuck also occurs as a street name in the town.

TATOMUCK (Fairfield County, Connecticut, and Westchester County, New York). Tatomuck Road is located in the town of Pound Ridge, New York. It runs near one of the headwaters of what was identified as the Tatomuck River in the July 18, 1640, Indian deed to land in the present-day town of Greenwich, Connecticut (Hurd 1881:365–66). The freshwater upper portion of the stream is today known as the Rippowam River. Its tidewater reach is called Mill Brook. Tomac Road, Lane, and Cemetery also preserve the name in somewhat altered form on Old Greenwich Neck where the Mill Brook empties into Stamford Harbor, which itself was called Tomuck Bay during colonial times.

TEEDYUSKUNG (Monroe and Pike counties, Pennsylvania). Today, the memory of the Delaware Indian king Teedyuscung is commemorated in the region as the namesake of Big and Little Teedyuskung lakes and the creek (also known as West Falls Creek) that joins the two together and carries their waters into the Lackawaxen River. Both are located in Lackawaxen Township, where Boy Scouts of America founder Dan Beard opened his Outdoor School in 1916 on land that he maintained on present-day Big Teedyuskung Lake as a private hunting and fishing camp. Teedyuscung was a very well connected, influential sachem and frontier diplomat and the successor of Nutimus, the central New Jersey leader who became the principal sachem of the Forks Indians at the time of the Walking Purchase. He played a leading role in his people's efforts to obtain redress for lands taken by the Walking Purchase and other tracts taken without payment in New Jersey. Failing to get compensation, he led his warriors against the British when the last French and Indian War broke out in 1755. Teedyuscung subsequently played a prominent part in the Easton treaty conferences held between 1757 and 1762 that restored the peace between his people and the British and settled outstanding accounts in New Jersey. A supporter of the Pennsylvania proprietors who built cabins for his followers at Wyoming when the Pennamite-Yankee disputes broke out into open conflict, he was burned to death when still-unknown arsonists torched his cabin at Wyoming in 1763 (Wallace

1991). Teedyuscung's name had lain all but forgotten on the pages of old colonial documents for more than a century and a half when the owners of Big and Little Tink ponds renamed them and the stream that flows from them into the Lackawaxen River in his honor during the 1930s.

TITICUS (Fairfield County, Connecticut, and Westchester County, New York). Whritenour finds that Titicus sounds like a Munsee word, **thihtukwus*, "cold little river." The name first appeared under its current spelling in Indian deeds to land in the area signed between 1715 and 1729 (Robert Bolton 1881[1]:393–94; Hurd 1881:635–38). Today, Titicus is the name of a dam and reservoir in New York, a river flowing into the state from Connecticut, mountains and roads in both states, and a hamlet in Connecticut. The nine-mile-long Titicus River rises in the Titicus section of the town of Ridgefield whose historic heart is preserved within the Titicus Historic District, listed in the National Register of Historic Places in 1985. The river then flows through a region of streams and ponds that includes Mopus Brook and Mamanasco Lake before flowing on into the New York City Water Supply System's Titicus Reservoir, first built in 1893. The river then runs another half mile through the hamlet of Purdys into the Croton River just above the Muscoot Reservoir. Titicus Mountain in Connecticut is a 1,026-foot-high hill located in the town of New Fairfield nine miles north of the hamlet of Titicus. The 925-foot-high Titicus Mountain (known in colonial times as Tom Spring Mountain) in New York is located several miles farther south near Sal J. Prezioso Mountain Lakes Park.

TOBYHANNA (Carbon, Monroe, and Wayne counties, Pennsylvania). Nora Thompson Dean suggested a Southern Unami origin for Tobyhanna in the form of *tëpihane*, "cold water creek." Following Heckewelder, Whritenour thinks it sounds much more like a Northern Unami word, *topihanne*, "stream of alder trees, or a creek on the banks of which that shrub grows spontaneously." Tobyhanna is the name of a creek, a township, a lake, a state park, an army depot, and other places centering around the Monroe County townships of Tobyhanna and Coolbaugh. The name first appeared as Tobyhannah Creek in William Scull's 1770 map. Tobyhanna Creek is a thirty-mile-long stream that rises in a swampy valley just a mile or so below the low divide that separates it from the headwaters of the Lehigh River at the southern end of Wayne County. The stream then flows south to Coolbaugh Township into Tobyhanna Lake located in the 5,440-acre Tobyhanna State Park. The park was opened in 1949 on land that had been an artillery firing range in the Tobyhanna Military Reservation. Established in 1912 and briefly closed following the end of World War II, the Tobyhanna Military Reservation was reopened in 1953 as the Tobyhanna Army

Depot, which presently functions as an electronic signals systems center. After flowing through the depot, Tobyhanna Creek passes through the village of Tobyhanna into state game lands in Tobyhanna Township. Entering Pocono Lake, the creek's waters pass through the Pocono Lake Dam gates into a stream bed that forms the boundary between Monroe and Carbon counties from Blakeslee to the brook's junction with the Lehigh River at Acahela.

TOHICKON (Bucks County, Pennsylvania). Heckewelder thought Tohickon sounded much like the Delaware words *tohíckhan* or *tohickhanne*, "the stream over which we pass by means of a bridge of drift wood." Today, thirty-mile-long Tohickon Creek flows through several townships from its headwaters above Quakertown into Lake Nockamixon. From there it runs below the steep red shale cliffs at High Rocks in Tohickon Valley County Park to its junction with the Delaware River at the village of Point Pleasant. The area was first mentioned by William Penn in a letter written on August 7, 1699, complaining that his surveyor had not measured the "Indian township called Touhicken, rich lands and much cleared by the Indians" (Dunn et al. 1981–86[3]:613). Several writers suggest that Penn's reference to the Indian township indicates that he intended to establish the kind of Indian reservation his descendants later briefly set aside at Indian Manor farther north in the Lehigh Valley. It seems more likely that Penn was much more anxious to acquire Touhicken's desirably rich and much cleared lands for himself and his children. Considerable evidence indicates that the Forks Indians thought that Tohickon Creek represented the northern bounds of the land they were expected to give up under the terms of the 1737 Walking Purchase deed. Penn's sons Richard and Thomas did not see it that way and used the results of the September 19, 1737, walk to claim land more than fifty miles north of the creek. Mills built along the Tohickon soon after the Indians were forced to leave the region quickly turned the creek into a working river. The district ultimately became a recreation area containing parks, such as the Tohickon Valley County Park, and camps, such as Camp Tohikanee, after the Great Depression of the 1930s forced the closings of the last mills operating along the creek. The name also appeared as Tohiccon at Tioga (present-day Athens, Pennsylvania) on the three versions of the Lewis Evans map published between 1749 and 1755 (Gipson 1939).

TOKONEKE (Fairfield County, Connecticut). Whritenour thinks that Tokoneke sounds very much like Taconic. Today, Tokoneke is the name of a road (Connecticut State Route 136), a park, and a neighborhood in the village of Darien. The name is a modern-day, slightly respelled resurrection drawn from records mentioning Tokaneke as one of the Indians who signed a deed to land in the area dated February 26, 1640 (in Robert Bolton 1881[1]:389–90).

TOMAC. See **TATOMUCK**

TOPANEMUS (Monmouth County, New Jersey). Whritenour (in Boyd 2005:443) suggests the translation "a lot of fish," from the Munsee words *tohpi*, "a lot of," and *namèes*, "fish." Rementer in the same source suggests *tèpi nàmès*, "enough fish," from the Delaware trade jargon. Today, Topanemus is the name of a six-mile-long tributary of Matchaponix Brook, a dam and lake built in 1915, and several other places in Marlboro and Manalapan townships. Colonists first mentioned an Indian town just west of the present-day village of Marlboro they called Toponemose in August 24, 1674 (New Jersey Archives, Liber 1:271[68]–270[69]), and Toponemes on June 25, 1696 (Budke 1975a:65A–65B). The Indians of Toponemus continued to live in their town for more than half a century after they sold their lands around it. They finally surrendered their claim to the town as part of the settlement hammered out at the Treaty of Easton in 1758. Most of the community's population moved to the Brotherton Indian Reservation established under the terms of the treaty, where they remained until 1801 when most moved to the Oneida Reservation in upstate New York. Colonists built a village they came to call Bucktown near Topanemus west of the present-day village of Marlboro in 1685. A Quaker meetinghouse was established at this locale in 1692. Residents had already changed the name of the place to Marlboro when the first post office was opened in the village in 1840.

TOQUAM. See **RIPPOWAM**

TOTOWA (Passaic County, New Jersey). Heckewelder thought that Totowa sounded like a Delaware word, *totauwéi*, "to sink, dive, going under water by pressure, or forced under by weight of the water." Today, Totowa is the name of a borough erected in 1898 as well as several other nearby places. The name first appeared as "Totoa on Pissaick River" in a November 3, 1696, confirmation of an earlier deed to land in the area (Whitehead et al. 1880–1931[21]:250). Many writers think that Totowa was the Indian name of the Passaic Falls in present-day Paterson. Goffle Mountain west of the borough of Totowa was also known as Totoway Mountain (Gordon 1834:8). Long a farming community, Totowa borough now contains a mix of homes and light industry.

TOTT. See **TATAMY**

TOWACO (Morris County, New Jersey). The origin of the name of the present-day hamlet of Towaco is not clear. It may come from the nearby 885-foot-high Waughaw Mountain, earlier known as Ta Waughaw, to the north of Towaco. It also may come from the 478-foot-high Towackhaw Mountain located just south of the hamlet. Both mountains probably bear the name of the prominent colonial Indian interpreter and intermediary Towackhachi (other spellings include Towwecoo and Towekwa), known among the settlers as Claes de Wilt or Claes the Indian. Like several other notable Indian diplomats in northern New Jersey, Towackhachi was originally from the east side of the Hudson River. Usually identified as Claes, he played a major role in land sales and other negotiations on both sides of the lower Hudson between 1666 and 1714. Suggesting that Towaco sounds very much like a Munsee word, *tuweekw*, "mudpuppy," Whritenour adds that the restricted range of mudpuppies (also called hellbenders or waterdogs) in the Hudson Valley limiting them to the east side of the river provides further evidence supporting the possibility that Claes the Indian was originally from there.

The present-day village of Towaco was referred to as Towagham as early as 1797. The locale was later renamed Whitehall for the station built there by the Delaware, Lackawanna, and Western Railroad during the late 1800s. The railroad changed the station's name to Towaco in 1905. Today, Towaco is a suburban community in Montville Township at the junction of U.S. Route 202 and Interstate 287. It is further linked to the metropolitan area by New Jersey Transit's Towaco Station on the Montclair-Boonton commuter rail line.

TUNKHANNOCK (Monroe County, Pennsylvania). Heckewelder thought that Tunkhannock closely resembled *tankhánne*, a Delaware word referring to a place where a smaller stream flows into a larger one. Whritenour thinks the name simply means "little stream" in Munsee. In Greater New York the place name Tunkhannock presently adorns a twenty-mile-long creek and one of the townships through which it flows in the Lehigh Valley. The name first appeared in present-day Monroe County as Tunkhana in Reading Howell's map of 1792. Tunkhannock Township was incorporated in the county by 1860. Today, the area embraced by places bearing the name Tunkhannock in the Poconos are mostly associated with small rural or suburban residential and resort communities. The identically spelled name of present Tunkhannock borough and Tunkhannock Creek in the Susquehanna River Valley were both first identified as Tankonink on William Scull's 1770 map.

TUXEDO (Orange County, New York). Noting that spellings of the place name Tuxedo preserved in colonial records resemble a Munsee animate noun referring to members of the wolf phratry, Whritenour thinks Tuxedo sounds like *ptukwsiituw*, "there are wolf phratry members." He goes on to point out that the root word of the Munsee wolf phratry name *ptukwsiit*, "round-foot," generally refers to dogs, foxes, and bears as well as wolves. Perhaps the name refers to the original Tuxedo Pond's round, pawlike shape. Today, Tuxedo is the name of the town of Tuxedo, the village of Tuxedo Park, and an artificial lake and dam at the western end of the village. The name first appeared as Tucseto and Tuxseto in the 1735 Charles Clinton survey field book for Cheesecocks Patent land (Freeland 1898:13). Shortly thereafter, it appeared as the highly anglicized Duckcedar Pond and later as modern-day Round Island and Pond.

The area in and around the present-day town of Tuxedo was a center of the region's iron industry between the Revolutionary and Civil wars. The town itself was first incorporated as Southfields in 1863 at a time when local companies responding to increased demand for iron stimulated by the Civil War significantly increased production and payrolls. Much of Southfields was returned to the town of Monroe after population levels diminished to prewar levels following the restoration of peace in 1865. The whole of earlier Southfields was reconstituted and renamed Tuxedo in 1890, just four years after local landowner Pierre Lorillard IV created the exclusive self-sustaining Tuxedo Park community in 1886. The black-tie, tailless dinner jacket known as the tuxedo received its name after it was first seen being worn at the annual Tuxedo Park autumn ball sometime around the turn of the twentieth century. Communities across the country subsequently adopted the name in the hope that some of its glamour would rub off on them. The original New York community of Tuxedo Park incorporated itself as a freestanding village in 1952 and was listed in the National Register of Historic Places in 1980. A significant portion of the present-day town of Tuxedo surrounding Tuxedo Park lies within the Harriman and Sterling Forest state parks.

UMPAWAUG (Fairfield County, Connecticut). Umpawaug Pond in the town of Redding is the source of the Saugatuck River. The name also appears on local maps as the name of a road, a school, and a cemetery. Umpawaug was first mentioned in Indian deeds to land in the area executed in 1680 and 1681 (Fairfield Town Records, Deed Book A:363, 417). It next appeared as a mile-square-sized tract called Umpawaug in a December 26, 1686, Indian deed in northern Fairfield and as Ompaquag in a conveyance dated September 12, 1687 (in Wojciechowski 1985:111). The name was first applied to the present-day pond in the form of Umpewange in a September 30, 1708, deed in the locale (in Robert Bolton 1881[1]:329–33). Umpawaug Hill was first noted in a colonial patent dated May 1, 1723. The Umpawaug District School, which opened in 1790 and has been closed since 1931, was listed in the National Register of Historic Places in 1988.

WAACKAAK (Monmouth County, New Jersey). Whritenour thinks that Waackaak, often phonetically spelled Waycake, sounds very similar to a Munsee word, *waakeek*, "that which is curved or bent." Waackaak Creek is a two-mile-long tidal stream whose course runs from its junction with Mahoras Creek at Philips Mills north to the place where it flows into Raritan Bay at the borough of Keansburg. The name was first mentioned as "Waycack upon the sea coast" in the boundary description written into the June 5, 1665, Indian deed to land at Navesink (Municipal Archives of New York City, Gravesend Town Records:74). Local residents continued to refer to their community as Waycake until Thomas Tanner erected a pier at the creek mouth known until the early 1820s as Tanner's Landing. Local tradition holds that residents changed the place's name around this time to Granville as a gesture expressing their appreciation of the Philips Mills grain-grinding capacity. Community residents finally adopted the name Keansburg in 1884 in honor of the prominent Kean family. Tourism joined fishing as primary engines fueling the village of Keansburg's growth into a formally erected borough in 1917.

WACCABUC (Westchester County, New York). Whritenour thinks the early form of the name, Wepack, sounds like a Munsee word, *xwupeek*, "it is a lot of water." Today, Waccabuc is the name of a hamlet, a lake, a country club, a three-mile-long creek that flows from the lake to its junction with the Cross River, and roads named Waccabuc and Waccabus in the town of Lewisboro. The name first appeared in colonial records as "Wepack or Long Pond so called" in the July 4, 1727, Indian deed to land in the Connecticut town of Ridgefield (Hurd 1881:636–37). Local entrepreneur Martin R. Mead resurrected the name when he built his Waccabuc House Hotel on Long Pond (soon renamed Waccabak Lake) in 1860. The name was adopted by the post office built nearby during the 1870s and by the country club established in 1912 on the site of Mead's hotel, which had burned in 1896.

WAGARAW (Passaic County, New Jersey). Wagaraw is a street name in the villages of Prospect Park, Hawthorne, and Fair Lawn just north of the city of Paterson. The name first appeared in a December 10, 1696, deed confirmation to land "on Pissack River below the mouth of Wachra Brook"

(Whitehead et al. 1880–1931[21]:248). Wagaraw also appeared as a road name on colonial maps.

WAHACKME (Fairfield County, Connecticut). Wahackme Road and Lane are street names in the town of New Canaan. Wahackme is a somewhat altered spelling of the name of a sachem identified as Mahackemo in the February 26, 1640 Indian deed to land at Norwalk and as Mahackem two months later in a deed to an adjacent tract (in Robert Bolton 1881[1]:389–90).

WALLENPAUPACK (Pike and Wayne counties, Pennsylvania). Heckewelder thought that Wallenpaupack sounded very much like a Delaware word, *wahlinkpapeek*, "deep and dead water." Today, Wallenpaupack is the name of a lake, a creek, the Wallenpaupeck Ledges Natural Area in the Lacawac Sanctuary, and a number of other places in the area. The name also occurs locally in modified forms as the name of Paupackan Lake, Paupack Township, the village of Paupack in Palmyra Township, and Lake Paupack in Greene Township. Wallenpaupack first appeared on William Scull's 1770 map as the "Wallenpanpack Branch of the Lechawaxin Creek." Present-day Hawley below the Lake Wallenpaupack Dam, built along the lower part of the creek in 1926 as a hydroelectric project by Pennsylvania Power and Light, was initially called Paupack Eddy. The thirteen-mile-long Lake Wallenpaupack Reservoir quickly became the focal point of the region's resort industry. The name also adorns the hamlet of Wallenpaupack Mills in the Wayne County township of Salem.

WALPACK (Sussex and Warren counties, New Jersey; Monroe and Pike counties, Pennsylvania). Whritenour thinks it highly probable that **waalpeekw*, "turn-hole, i.e., whirlpool or water hole," is a Munsee cognate of the Delaware word *walpeek*, which Heckewelder translated as "a turn hole, a deep and still place in a stream." The distinctive elongated S-shaped curve of Walpack Bend on the Delaware River has made it a readily recognized boundary marker for the four counties in two states that meet there. Walpake was first noted in 1731 as the name of colonial settlements located on both sides of the Delaware River. Walpack on the New Jersey side was designated as a precinct one year after Sussex County was erected in 1753. The settlement became the site of the stockaded Fort Walpack in 1756 during the final French and Indian War and became the center of Walpack Township (incorporated in 1798). Today, the original sites of the Walpake communities both lie within the Delaware Water Gap National Recreation Area. The name also marks the 388-acre Walpack Wildlife Management Area managed by the New Jersey Department of Environmental Protection, as well as a number of other places in the area.

WAMPUS (Fairfield County, Connecticut, and Westchester County, New York). In New York, Wampus is the name of a lake, a river, a park, a school, and several other places located in and around the hamlet of Armonk in the town of North Castle. The name was probably fixed onto Westchester maps sometime before Robert Bolton (1881[1]:362–63) published a transcript of the October 19, 1696, deed to land in the area signed by a sachem identified as Wampus in the first edition of his history in 1848. The name also adorns Wampus Way in the Lake Hills development built in the Connecticut town of Fairfield in 1952. Wampus was almost certainly Wampage, a local sachem also identified as Wampasum and Wampegon in deeds documenting land sales in the area between 1651 and 1696. His other documented name was Ann Hook, a sobriquet long thought to be a trophy name he used to commemorate his killing of Anne Hutchinson in 1642. While the name Ann Hook invoked Hutchinson's memory, it was more likely taken from Anne's Hoeck, a neck of land jutting into Eastchester Bay evidently named for Hutchinson. Later called Pell's Point and now known as Rodman's Neck, the place is the current site of the New York City Police Department training facility and firing range in Pelham Bay Park.

Wampage's connections, if any, with John Wampus, the central Massachusetts Nipmuck émigré who claimed land in modern-day Fairfield County, Connecticut, in 1671, currently are unclear. Although town fathers in Westchester had the Wampus who signed the 1696 deed in mind when they chose the name for their community during the 1840s, the particular spelling they selected also evokes the image of the Wampus cat, an infernal feline that still stalks the woods in Cherokee and Appalachian Mountain folktales. Wampus remains a popular name on Westchester maps. The Westchester County Department of Parks, Recreation, and Conservation manages Wampus Pond and ninety-three acres of shoreline fronting the lake acquired from the City of New York in 1963.

WANAMASSA (Monmouth County, New Jersey). Wanamassa was first mentioned as one of the three sachems signing over land at the head of present-day Deal Lake on April 6, 1687 (New Jersey Archives, Liber D:147–49). It was resurrected as the name for a YMCA camp at the current Wanamassa locale just west of Asbury Park in 1892. The name subsequently came to adorn a bungalow colony, a hotel, and, finally, the surrounding neighborhood in modern-day Ocean Township.

WANAQUE (Passaic County, New Jersey, and Rockland County, New York). Whritenour thinks Wyanokie, a form of the name (locally pronounced win-o-key) that shares spaces on area maps with Wanaque (often pronounced wah-na-cue), sounds much like a Delaware word, *winakwi*, "sassafras." The name first appeared as Wanochke Brook in an Indian deed to land in the area dated September 8, 1729 (Roome 1897:24). Today, the name Wanaque adorns a reservoir and dam, a river, a township, a borough, and numerous other places at the northeastern end of Passaic County. Hikers have long referred to the uplands in and around the area as the Wyanokie Plateau. Since 1972 a consortium of five West Essex County townships has owned and managed Camp Wyanokie, a 150-acre camping and hiking preserve established on the shores of Boy Scout Lake in West Milford as a private camp in 1919. Another spelling of Wanaque adorns present-day Camp Winaki Road in New York's Harriman State Park. The fifteen-mile-long Wanaque River flows out of Greenwood Lake (known as Long Pond during the eighteenth century and the source of the Wanaque River's original name, Long Pond River) at the village of Awosting in West Milford Township. From there it runs into the 2,320-acre Wanaque Wildlife Management Area incorporated into Long Pond Ironworks State Park in 2009. The river then flows past the Long Pond Ironworks built in 1766 through the Lake Monksville Reservoir where it enters Ringwood Township. In Ringwood the stream flows into the five-mile-long Wanaque Reservoir (completed in 1928) that begins just below the southern end of the Lake Monksville Reservoir. Outflow from the Raymond Dam holding back the Wanaque Reservoir's waters runs into the Wanaque River as it flows through the borough of Wanaque. This lower section of the stream was known as the Ringwood River before the creation of the Wanaque Reservoir. The Wanaque River then flows south into the borough of Pompton Lakes, where its joins with the Pequannock River just a mile north of the place where the junction of the Pequannock and Ramapo rivers forms the Pompton River.

The name's other spelling first adorned the Wyanokie Mine, the Wyanokie Furnace, and the hamlet where worker housing was built in the Ringwood Valley area by the 1860s. A village called Midvale was established at the site of the Wyanokie Furnace around the station that served as the terminus of the Montclair and Greenwood Railroad, which was extended in 1872 to serve the ironworks along the Ringwood River. Midvale residents adopted the current spelling of the furnace's Indian name when they joined with the village of Haskell to form the borough of Wanaque in 1918 (Wardell 2009:112).

WAPETUCK (Westchester County, New York). Whritenour thinks Wapetuck sounds much like a Munsee word, *waapihtukw*, "white river." Today, Wapetuck is the name of a municipal school board day camp operated by the village of Scarsdale. Wapetuck was originally the name of one of the Indians who signed four deeds to land in the area between 1701 and 1702 (in Robert

Bolton 1881[1]:475–76; [2]:211–12; New York State Library, Indorsed Land Papers [3]:33).

WAPPINGER (Dutchess County, New York). Whritenour thinks Wappinger sounds a great deal like a Munsee word, **waapingw*, "white face, i.e., opossum." This conforms with current opinion casting doubt on earlier etymologies of "easterner or dawnlander." The name's similarity to the London dockyard district of Wapping (Saxon for "Waeppa's people") doubtless eased its transition onto colonial maps. Today, Wappinger is the name of the thirty-six-mile-long creek that courses through the heart of Dutchess County, its East and Little Branch tributaries, and a town, a lake, and the village and waterfall of Wappingers Falls. People identified as Wappinger Indians appeared early and often in records documenting intercultural relations along the Lower Hudson River's Great Valley during the colonial era. A reference to Wappings living on the North River (today's Hudson) halfway between forts Orange and Amsterdam made in 1643 during Governor Kieft's War represents the earliest known appearance of the name. Although a 1653 reference (in Grumet 1994) mentioning the place name Opingua at today's Ramapo Pass is the first occurrence of the name Oping on the New York–New Jersey frontier, Indians variously calling themselves Wapings, Opings, and Pomptons only began being mentioned in documents in that area some twenty-five years later. While the names Wappinger and Waping occur in the Great Valley on both sides of the Hudson River, no source dating to colonial times mentions a Wappinger confederacy dominated by a Wappinger chieftaincy first proposed by Ruttenber (1872:77–85). The name remained on Dutchess County maps as the name of Wappinger Creek after the Wappinger Indians left the area following their sachem Daniel Nimham's failure to regain title to his people's land there in 1765. Today's village of Wappingers Falls was known as Wappingers Creek from late colonial times to the early federal period. The entire locale was known as Channingville from 1847 until 1849, when part of the community split off to form the hamlet of Wappingers Falls (Kaiser 1965). Both places were united into a single community in 1871, when residents drew up articles of incorporation for a single village they named Wappingers Falls.

WARACKAMACK (Dutchess County, New York). Golden (2009:11–12) thinks that the name Wohnockkanmeekkuk first mentioned in a "Muhheckaunuck or River Indian" petition dated June 29, 1754 (New York Colonial Manuscripts, Indorsed Land Papers [15]:283–84), was their name for present-day Warackamack Lake just outside of the village of Red Hook. The spelling adopted by the builder of the dam flooding a stretch of marshlands known as Fever Cot or Pine Swamp indicates that it was drawn from Ruttenber's (1906a:46) listing of Waraughkameck in his place name book.

WARAMAUG (Litchfield County, Connecticut). Waramaug is presently the name of a lake, a brook, a state park, and a country club. Lake Waramaug is a natural body of water deepened and enlarged by a dam where the borders of the towns of Kent, Washington, and Warren meet. The lake is fed by Lake Waramaug Brook (also called Sucker Brook) and several other small streams. The Waramaug Country Club first opened as a golf course in 1893. Ninety-five-acre Lake Waramaug State Park is managed by the Connecticut Department of Energy and Environmental Protection on land purchased by the state in 1920 on the lake's north shore. The East Aspetuck Creek carries water from the lake to the Housatonic River at New Milford. The name Waramaug originally appeared in colonial records documenting the participation of a sachem variously identified as Weramaug and Weromaug in land sales in the area negotiated between 1716 and 1720 (in Wojciechowski 1985:138–40).

WARINANCO (Union County, New Jersey). Warinanco is a spelling that a colonial scribe used to identify one of the three sachems who signed the October 28, 1664, Elizabethtown deed transcribed in Whitehead et al. (1880–1931[1]:15–16). The full name of this sachem appears elsewhere as Waerhinnis Couwee, a local Indian leader who participated in land sales concluded in the area between 1639 and 1677. Union County Park Commission officials selected the name for their Warinanco Park built in the communities of Elizabeth and Roselle in 1925.

WASSAIC (Dutchess County, New York). Wassaic Creek is a six-mile-long stream rising above the village of Amenia that flows south to its junction with Webatuck Creek to form the Tenmile River a mile or so above the village of Dover Plains. The name first appeared as Wesaick Brook in the February 1704 survey of land located in what was called the Oblong tract between Connecticut and New York conveyed by the Indians in a deed dated November 5, 1703 (O'Callaghan 1864:74–75). Colonists often interchangeably used the names Weebatuck and Wassaic when referring to the present-day Tenmile River. The Wassaic Creek was also identified as Steel Works Creek in Spafford's gazetteer (1813:124). The hamlet of Wassaic grew up at its present locale at the headwaters of the Wassaic Creek in the village of Amenia during the mid-eighteenth century. More intensive development began after a railroad station opened at the village by 1860 (French 1860:270). Demand for condensed milk produced at the Borden plant in Wassaic stimulated by the Civil War brought the locale to national attention. Today, the name Wassaic continues to adorn the village as well as the Wassaic Creek and the 488-acre Wassaic Multiple Use Area located on the Tenmile River just below Wassaic Creek's confluence with Webatuck Creek.

WATCHUNG (Essex, Middlesex, Passaic, Somerset, and Union counties, New Jersey). Nora Thompson Dean thought the name Watchung resembled a Southern Unami word, *ohchung*, "hilly place." Whritenour agrees, offering a Munsee cognate, *wahchung*, "in the hills." The name presently adorns a substantial number of places in and around the Watchung Mountains that rise above coastal plain valleys drained by the Passaic, Rahway, and Raritan rivers. The name first appeared as Watchung Mountain in the July 11, 1667, Indian deed to land at Newark (New Jersey Archives, Liber 1:270[69]). Today, the name most notably adorns the three ranges of the Watchung Mountains, Watchung Township in Union County, and the two-thousand-acre Watchung Reservation built as a flagship facility by Union County Park Commissioners, who began acquiring land for the park in 1921. The Watchung Reservation became one of several parks built on more than four thousand acres purchased by commissioners by 1930. The commission was dissolved in 1978 and its lands, which had by then increased in size to 6,700 acres, were placed under the management of the Union County Department of Parks and Recreation.

WATNONG (Morris County, New Jersey). Whritenour thinks Watnong sounds like a Munsee word, **xwahtunung*, "at the big mountain." Today, Watnong is the name of a mountain, a brook, and other places in and around the village of Morris Plains. The name first appeared as "an Indian plantation called Whattanung" in the June 1, 1716, entry in John Reading, Jr.'s (1915:46) survey book. Present-day Morris Plains was known as the Watnong Plains during colonial times. Six-mile-long Watnong Brook runs from its headwaters below Union Hill in Denville Township into Parsippany–Troy Hills Township through the village of Tabor, past 965-foot-high Watnong Mountain, and into the village of Morris Plains, where it joins with the Whippany River.

WATSESSING (Essex County, New Jersey). Watsessing is presently the name of a neighborhood, a park, a railroad station, a hill, and a street in the townships of Bloomfield and West Orange. The name first appeared on two patents to lands in Newark made in 1696, the first to Watsesson's Plain on April 27 (Whitehead et al. 1880–1931[21]:244) and the second to Watsesons Hill on December 8 (ibid.:256). The community of Wardsesson grew up as a mill town along the upper branches of the Second River, a Passaic River tributary, just above Watsessing Dock. Residents changed the name of their village to Bloomfield in 1799 and subsequently kept it for their township after they split off from Newark in 1812 (Folsom 1912:43–44). Watsessing currently adorns the Watsessing and Watsessing Heights neighborhoods, Watsessing Avenue, the Watsessing Railroad Station (built in 1912), the Watsessing Post Office, and Watsessing Park, an Olmstead-designed recreational facility in Bloomfield and West Orange first commissioned by Essex County in 1899.

WAUGHAW (Morris County, New Jersey). The origin of the name of 885-foot-high Waughaw Mountain and nearby Waughaw Road is unclear. It may come from the Wachra Brook, first mentioned in a December 10, 1696, deed confirmation to land "on Pissack River" (Whitehead et al. 1880–1931[21]:248) and preserved on subsequent maps as Wagaraw. An early spelling of the mountain's name as Ta Waughaw suggests that it also may be a rendering of Towackhachi, namesake of the nearby Towaco.

WAUGHKONK (Ulster County, New York). Waughkonk Road in the town of Kingston marks the memory of the name of a place first noted as Anguagekonk in a March 12, 1702, petition to purchase three hundred acres of land on the nearby Saw Kill (New York State Library, Indorsed Land Papers [3]:40). Esopus leaders attending a Nicolls Treaty renewal meeting twenty years later in August 12, 1722 (Philhower Collection AC 1810, Alexander Library, Rutgers University), complained that they had not been paid for land at places they identified as Ashewagkomek (noted as near Mammekotten [Mamakating] and Nepanagh [Napanoch]) and at Waghkonk, another locale noted as closer to present-day Waughkonk Road.

WAWARSING (Sullivan County, New York). Whritenour thinks Wawarsing sounds similar to a Munsee word, *wahwalusung*, "place of little eggs." The name first appeared in a January 3, 1672, report as Waewaersinck, the destination of four people whom local settlers identified as otherwise unspecified Southern Indians (in Fried 2005:13–14). It stayed on local maps as the name of the colonial community established at the current locale and was given to the present-day town of Wawarsing when it was formed from land split off from the town of Rochester in 1806. The village of Wawarsing grew large enough to support a post office by 1817 (Kaiser 1965). Wawarsing became a depot town situated along the route of the Delaware and Hudson Canal and Railroad during the nineteenth century and a resort town during the twentieth. Today it is a modest residential community joined to nearby Kerhonkson and Ellenville by U.S. Route 209.

WAWAYANDA (Sussex and Passaic counties, New Jersey; Orange County, New York). Whritenour thinks Wawayanda sounds like a Munsee word, *wehwalhandi*, "the ditch." Local folk etymologists have long maintained that the name is a transcribed version of the expression "way, way, yonder." Today, a mountain, two lakes (Wawayanda and New Wawayanda), a creek that flows across the state line between New York and New Jersey, and a state park in New Jersey most notably bear the name. Wawayanda was first mentioned in a conveyance dated March 5, 1703 (Budke 1975a:79–81), and was shortly thereafter noted as the name of "a creek called Wawajando" in another deed signed on March 30, 1703 (New York State Library, Indorsed Land Papers [3]:177). It may also be the place identified as Waweyaghponekan in an Indian deed to land in the area signed on April 26, 1712 (National Museum of the American Indian, Specimen No. 24/6665). John Reading, Jr. (1915:91, 109), referred to the area as Oweonda, "a place mightily stored with meadow," on June 4, 1716, and as the Waweonda Drowned Lands on July 27, 1719.

The seven-mile-long main stem of Wawayanda Creek rises in Wickham Lake in the town of Warwick. It then flows west where it is joined by a creek flowing from Lake Wawayanda in the 34,350-acre Wawayanda State Park, managed by the New Jersey Division of Parks and Forestry. From there it flows into New Jersey along the base of the northwest scarp of Wawayanda Mountain to its junction with Black Creek in the Vernon Valley. The combined waters of the stream, known as Pohuck Creek, flow north back into New York where they join with the Wallkill River at Pochuck Neck in the heart of the Black Dirt District. Managers of Lake Aeroflex in Kittatiny Valley State Park renamed the pond (whose 101-foot depth makes it the deepest natural body of water in New Jersey) New Wawayanda Lake when an expansion project joined the original holder of the name to Lake Wawayanda.

WEAMACONK (Monmouth County, New Jersey). Lucy Parks Blalock (in Boyd 2005) thought Weamaconk sounded very much like a Southern Unami word, *wémakung*, "place where there is almost nothing but trees." Today, the six-mile-long Weamaconk Creek rises in the borough of Freehold, where it flows west through the Monmouth Battlefield State Park to the place where it is joined by the three-mile-long Weamaconk Brook. From there the creek flows past the village of Tennent to its junction with McGellairds Brook at Englishtown. Known below the junction as the Matchaponix Brook, the creek's waters ultimately flow into the Manalapan Brook and on through the South River into Raritan Bay at Keyport. Wemcock Point where the Weamaconk Creek and Brook meet was first noted in a document dated June 30, 1691 (Whitehead et al. 1880–1931[21]:149). It next appeared as Wemcoke in a deed signed on May 15, 1700 (ibid.:142) and as Wem-cook Point in the August 1, 1716, Indian deed to land later renamed Davison Neck (Monmouth County Records, Deed Book E:197). Revolutionary War history buffs will recognize the form Wemrock Creek as the name of the stream where Molly Pitcher drew water to swab down the artillery piece she served when her husband fell wounded next to the gun during the Battle of Monmouth on June 28, 1778.

WEBATUCK (Litchfield County, Connecticut, and Dutchess County, New York). Whritenour thinks Webatuck sounds like a Munsee word, *wiipihtukw*, "arrow tree" or "arrow river." He further notes that the earliest spelling of the name, Weputing, resembles *wiipahtung*, "arrow mountain." Ten-mile-long Webatuck Creek rises in Sharon, Connecticut, and flows across the state line into New York just east of the village of Millerton. From there it weaves in and out across the state line until it falls into the Tenmile River at its junction with Wassaic Creek just below South Amenia. The name first appeared as a mountain called Weputing overlooking present-day Wassaic Creek mentioned in the February 1704 survey of land sold by the Indians on November 5, 1703, in the Oblong tract between Connecticut and New York (O'Callaghan 1864:74–75). The spelling of the name changed from Weputing to Webatuck as it traveled downhill from the mountain to the creek. Webatuck Creek was noted as Oblong or Weebotuck Creek in a gazetteer published by Spafford (1813:124). Local residents often used the names Webatuck and Wassaic interchangeably when referring to the entire present-day Webatuck-Tenmile river system (Reed 1875:10).

The present-day Webatuck locale began to draw people from New York City looking to escape the rigors of urban life soon after the New York and Harlem Railroad extended its direct line from Manhattan to Dover Plains in 1848. Amicable relations between local residents and city folk ultimately created a relaxed social climate that drew artists, writers, and other creative types looking for affordable country accommodations in a tolerant community. In 1926 Lockwood Hunt and his brother began producing handcrafted furniture in a building in Wingdale that became the nucleus of today's Webatuck Craft Village. One year later, Left-leaning progressive labor activists built what they billed as the world's first interracial adult summer camp at a place nearby they named Camp Unity. During the 1940s the camp became a haven for battle-worn American volunteers returning from service in the

Spanish Civil War. In 1958 the Camp Unity community renamed its facility Camp Webatuck to mark its shift from an adult retreat to a childrens' summer camp. Several of the children who attended Camp Webatuck between 1958 and 1966 grew up to be artists, writers, and musicians. Today, the name Webatuck continues to adorn its namesake creek, the craft village near its banks, and numerous street and business signs in the area.

WEEHAWKEN (Hudson County, New Jersey, and New York County, New York). Whritenour thinks Weehawken sounds like a Munsee word, *xwiiahking*, "at the big land." Today, Weehawken is the name of a township on the New Jersey side of the Hudson River and a street on Manhattan where the slip for the ferry running from New Jersey was located before completion of the Holland Tunnel in 1927 put an end to its service. The name was first mentioned as a "great clip [cliff] above Wiehacken" in an Indian deed to land in the area dated January 30, 1658 (New Jersey Archives, Liber 1:3–6). Colonial settlements built at the locale grew around the ferry that began running from Weehawken to Manhattan during the early 1700s. Local residents subsequently incorporated their community as Weehawken Township in 1859 (Wardell 2009:114).

WEEQUAHIC (Essex County, New Jersey). Whritenour has found that Weequahic sounds much like a Munsee word, *wihkweek*, "that which is the end of something (i.e., the head of a stream)." Today, Weequahic Lake is located in 311-acre Weequahic Park in the city of Newark. The name comes from Weequachick, the "great creek" marking the south bounds of the July 11, 1667, Indian deed to Newark (New Jersey Archives, Liber 1:270[69]). Colonists called the creek Branch Brook. The broad, flat fields of the Waverly Fairgrounds that opened at the locale in 1866 made it an ideal site for state fairs and horse races. This all came to an end when Essex County Park Commissioners acquired the land in 1895. Resurrecting the old Indian name Weequahic, they applied it to the park and public golf course (the oldest in the United States) completed by the Olmstead Brothers landscaping firm a few years later. The name was subsequently expanded to include the residential neighborhood that grew up alongside the park during the early 1900s. Designated as the Weequahic Park Historic District, the neighborhood was listed in the National Register of Historic Places in 2003.

WEMROCK. See **WEAMACOCK**

WERIMUS (Bergen County, New Jersey). Whritenour thinks the earliest orthography of the name, Weromensa, sounds like a Munsee word, *xweelameenzuw*, "there are many little fish." Today, Werimus is the name of a road that runs through the communities of Saddle River, Woodcliff Lake, and Hillsdale. Its first occurrence in colonial records, as Weromensa, an "old Indian field or plantation" located to the east of the Saddle River, was noted in an Indian deed to land in the area dated June 1, 1702 (Budke 1975a:84–86). The place later appeared as Awessawas Plantation (Wardell 2009:114).

WESTCOLANG (Pike County, Pennsylvania). The name of present-day Westcolang Creek and Pond is a slightly modified spelling of Wescolong, one of the Ninnepauues or Delaware Indians who signed over land between the Delaware and Susquehanna rivers to New England purchasers in May 6, 1755 (Boyd and Taylor 1930–71[1]:308–14). Today, Westcolang is the name of a two-mile-long creek that flows into the Delaware River below the dam holding back the waters of Westcolang Pond in the hamlet of Westcolang in Lackawaxen Township.

WHIPPANY (Morris County, New Jersey). Heckewelder stated that Whippany reminded him of a Delaware word, *wiphanne*, "arrow creek, where the wood or willow grows of which arrows are made." Whritenour points out that the name sounds more like a Munsee word, **xwihpunung*, "place of big tubers," combining *xw*, "big," and *ihpunung*, "place of tubers." The name first appeared as a "South Branch of the Passaieck River alias Monepening" (Whritenour suggests a variant of "tuber place" prefaced by *men*, "collected together in one place") in a March 10, 1690, Indian deed to land in the area (New Jersey Archives, Liber K-Large:170). It was next mentioned as the Machiponing River [another "tuber place" variant, this one modified by the word *meex*, "large" or *maxk*, "red"] "at the west side of the south branch of Pessyack River" in a deed dated December 3, 1701 (ibid., Liber O:145–48). Other spellings of the river's name in early colonial records include Mochwihponing on July 29, 1702 (New Jersey Historical Society, West Jersey Historic Manuscript Group 3:16), Mechwikponing on July 29, 1702 (New Jersey Archives, Liber M:555–56), Weypenunk on August 13, 1708 (ibid., Liber I:210–11), and Whippaning in John Reading, Jr.'s (1915:35) survey-book entry dated April 19, 1715. Today, the Whippany River and its branches flow from the city of Morristown past the village of Whippany in Hanover Township and along the border between East Hanover and Parsippany–Troy Hills townships to its junction with the Rockaway River just one mile from the latter river's confluence with the Passaic River at Hatfield Swamp.

WHITE PLAINS. See **QUAROPPAS**

WICCOPEE (Dutchess and Putnam counties, New York). Whritenour thinks Wiccopee sounds like Northern Unami words *wikbi*, "bast or inner bark," and **wihkwpi*, "end of the water (as in headwater)." Today, Wiccopee is the name of a creek and a hamlet in the Dutchess County town of East Fishkill and a pass, a reservoir, and a brook that flows south from Clarence Fahnestock Memorial State Park into the Peekskill Hollow Creek in Putnam County. The name first appeared in colonial records as a Fishkill Creek tributary identified as Wakapa Creek (see Mahopac) in a 1753 Philipse Patent surveyor's map (Library of Congress, Maps of North America, 1750–89:1083). It was later noted as a major Wappinger Indian settlement in present-day Dutchess County. Johannes Swartwout was the first colonist to begin work on a farmstead where Wiccopee Creek flows into the Fishkill. A descendant mentioned his ancestor's later repurchase of the tract from Wappinger Indians in a letter requesting title to the place sent to the superintendent of Indian affairs, Sir William Johnson, in 1762 (Sullivan et al. 1921–65[10]:493). The grateful descendant probably renamed the locale Johnsonville after the superintendent ruled against the Wappinger Indian claim to the area in 1765. A post office built at Johnsonville in 1826 bore the name until 1846 (Kaiser 1965). Local residents probably renamed the place Wiccopee sometime shortly thereafter.

WICKAPECKO (Monmouth County, New Jersey). Whritenour thinks Wickapecko sounds like a Munsee word, **wihkupeekw*, "the end of the pond." Today, Wickapecko is the relatively recently revived name given to a pond and a street in the hamlet of Wanamassa near the city of Asbury Park in Ocean Township. The name first appeared in an April 6, 1687, Indian deed to land in the area "within the branches of a great pond called by the Indians Whekaquecko" (New Jersey Archives, Liber D:147–48).

WICKATUNK (Monmouth County, New Jersey). Nora Thompson Dean (in Kraft and Kraft 1985:45) suggested "finishing place or end of a place (i.e., a trail)," from a Southern Unami word, *wikwètung*. Whritenour thinks the word could be a Northern Unami word, *wikhattenk*, "neighborhood," or—what he thinks is more likely—*wickatink*, "place of the leg or legs." Wickatunk is the name of a small community in Marlboro Township. The place was first noted as Weakatong in an Indian deed to land in the area dated June 5, 1665 (Municipal Archives of the City of New York, Gravesend Town Records:74). Colonists knew the area as a center of Indian settlement. Discussions among the East Jersey Board of Proprietors in 1685 concerning the purchase of the 36,000-acre Wickatunck tract resulted in the subsequent February 25, 1686, purchase (New Jersey Archives, Liber A:264) that included but did not mention Wickatunk (although it referred to the Indian town of Toponemes). Wickatunk presently is the name of a hamlet, a railroad station, a street, and several other places in the greater Marlboro area.

WICKECHEOKE (Hunterdon County, New Jersey). Fifteen-mile-long Wickecheoke Creek flows from its headwaters in the hills west of Flemington south through the hamlets of Croton and Locktown into a rocky stretch of gorges, rapids, and waterfalls above the Sergeantsville Bridge, New Jersey's last surviving covered span. The creek then flows alongside the Lower Creek Road to its junction with the Delaware River at Prallsville. Several locales in the stream's watershed are managed within the more than twenty-thousand-acre Wickecheoke Creek Preserve, the largest tract of its kind administered by the New Jersey Conservation Foundation. The name first appeared as Wickakicke Creek and Wickakick Brook in an Indian deed to land in the area signed on October 12, 1684 (New Jersey Archives, Liber A:262). Lewis Evans (in Gipson 1939) noted the stream as Watchoak Creek in his 1738 map of the Walking Purchase. The name appeared as Wickhechecoke Creek in a spelling similar to its modern form in Gordon's (1834:264) gazetteer.

WICKERS CREEK (Westchester County, New York). Wickers Creek is an anglicized version of Wiechquaesgeck, the original name of the Dobbs Ferry locale that colonists later used as a general term identifying all Indians living in and around Westchester. Whritenour thinks Wiechquaesgeck sounds very much like a Munsee word, *wihkwaskeekw*, "end of the swamp." Present-day Wickers Creek is a one-mile-long stream that flows through Dobbs Ferry into the Hudson River. The stream itself was first mentioned as "a creek or fall called by the Indians Weghqueghe and by the Christians called Lawrence's Plantation" in an Indian deed to land in the area dated April 13, 1682 (in Robert Bolton 1881[1]:269). An anglicized form of the name initially appeared in a will filed by landowner Frederick Philipse on October 26, 1700, which also referred to the stream as "a creek called by the Indians Wysquaqua and by the Christians as William Portuguese Creek" (in Pelletreau 1886:23–28). The Wiechquaesgeck Indians mostly moved to Nimham's Wappinger community following their last sales of their Westchester County lands around this time.

Today, Wickers Creek and the Wickers Creek Archaeological Site discovered near the stream's mouth in the 1980s carry on the name. The name also may be preserved by Wykagyl Country Club in a form, first documented on early maps, long thought to be a shortened version of Wiechquaesgeck. The club has operated at its present location in New Rochelle continuously since

1904. The architect Alfred Feltheimer, responsible for designing and naming stations along the interurban New York, Western, and Boston Railway on the north shore of Long Island Sound, gave the country club's name to the station built near it in 1912. The neighborhood of upscale houses that grew up around the station continues to be known as Wykagyl.

WILLOWEMOC (Sullivan County, New York). Willowemoc Creek and the hamlet of Willowemoc are located in Catskill State Park in the towns of Rockland and Neversink. The creek, which flows into the Beaver Kill at the village of Livingston Manor, is a nationally renowned trout fishing stream. Today, much of the land along its banks lies within the 14,800-acre Willowemoc Wild Forest Preserve managed by the New York Department of Environmental Conservation. Catskill poet Alfred B. Street (1845:49) based his invention of the currently used spelling of Willowemoc on earlier occurrences of the name in the forms of Whitenaughwemack on a 1785 survey map and Willowemock on the 1779 Sauthier map.

WINNIPAUK (Fairfield County, Connecticut). An Indian identified as Winnapucke first appeared in colonial records as one of the signatories to a deed conveying title to land in Norwalk to colonists on February 15, 1651 (Hurd 1881:483–84). This man was almost certainly Winnepoge, a brother of Nonopoge who, along with Craucreeco (also on regional maps as Cockenoe, Cricker, and Kensico), acceded to demands made on May 5, 1684, that they acknowledge colonial sovereignty over their lands as the price paid for their people's support of the Pequots, defeated by the English nearly fifty years earlier (J. Davis 1885:121–22). The sachem's name appeared in its present-day form in the deed of gift presenting the Norwalk Islands to a settler signed on December 2, 1690 (Selleck 1896:28). Today, Winnipauk is the name of a neighborhood, a millpond, and a street in the city of Norwalk. Winnepoge Road is one of the Indian names provided by the Fairfield Historical Society to the developer of the Lake Hills subdivision in 1952.

WOOSAMONSA (Mercer County, New Jersey). Whritenour thinks Wishilimensy, Achilomonsing, and similar early recorded forms of this place name sound like the Munsee words *wchulamiinzhuy*, "wrinkled or shriveled tree," or *wchulamiinzhiing*, "place of wrinkled or shriveled trees." Today, Woosamonsa is the name of a small country road in Hopewell Township. The route to the present-day name of Woosamonsa begins with Wishilimensy, a place near Hockin Creek (today's Alexauken Creek) mentioned in an April 16, 1701, entry for survey of three thousand acres of West Jersey Society land

(Whitehead et al. 1880–1931[21]:388). The name next appeared as Wighalimensy in a November 16, 1701, survey return for the tract (ibid.). An "Indian path leading from Itshilominwing unto Noshaning" near "a small brook or run having its rise or first spring about Achilomonsing" was next mentioned in an Indian deed to land nearby signed on June 5, 1703 (New Jersey Archives, Liber AAA:443–45). Various spellings of the name have subsequently been used as local road and school names (Hunter and Porter 1990). It was also used to identify modern-day 442-foot-high Pennington Mountain as recently as 1908.

WYANOKIE. See **WANAQUE**

WYANTENOCK (Litchfield County, Connecticut). The Wyantenock Indian towns located along the Housatonic River around New Milford were mixed communities made up of Wampano-speaking people from the area, Mahicans living farther north and east, and more westerly Munsees. Most of the inhabitants of these towns ultimately joined the nearby Schaghticoke Indian community by the second quarter of the eighteenth century. Today, the name most notably adorns the more than four-thousand-acre Wyantenock State Forest that extends across lands first acquired by the state of Connecticut in 1925 in the towns of Warren, Kent, and Cornwall. Slightly differently spelled Weantinock is a street name in the present-day village of New Milford.

WYKAGYL. See **WICKERS CREEK**

WYOMING (Luzerne County, Pennsylvania). Heckewelder wrote that Wyoming sounded like the Delaware words *m'cheuwómi* or *m'cheuwámi*, "extensive level flats." Whritenour seconds Goddard's findings indicating that the name is equivalent to a Munsee word, *xweewamung*, "at the big river flats." Wyoming Mountain is a low ridge midway between Moosic and Nescopeck Mountain situated along the divide separating the drainage of the Lehigh Valley from the Susquehanna River watershed. Major Munsee refugee settlements clustering around Wyomink in present-day Wilkes-Barre lay just to the west of the present-day Wyoming Mountain whose eastern slopes drain streams that now flow through the southwesternmost edges of the Greater New York area.

YANTACAW (Essex and Passaic counties, New Jersey). Whritenour suggests that Yantacaw sounds somewhat similar to a Munsee word, *yu wandakw*, "here this way." Today, Yantacaw is the name of a brook, a river, a pond, a park, several streets, and a number of other places in and around the cities of Montclair,

Bloomfield, and Nutley. The Yauntakah River was first mentioned in the July 11, 1667, Newark Indian purchase (New Jersey Archives, Liber 1:270[69]). Jandakagh later appeared as one of five named parcels in northern New Jersey conveyed in an Indian deed dated October 10, 1700 (Budke 1975a:77–78). The nine-mile-long Yantacaw River, also known as the Third River, rises at the Great Notch Reservoir in West Paterson Township. From there the stream flows south past Little Falls into the Essex County townships of Upper Montclair and Montclair, where it is joined by the one-mile-long Yantacaw Brook. It then passes through Glen Ridge, Bloomfield, and Nutley townships to its confluence with the Passaic River.

YAW PAW (Bergen County, New Jersey). Whritenour thinks that Yaw Paw sounds like a Munsee word, **yaapeewi*, "on the edge of the water." Yawpaw was first mentioned as Japough, another of the five parcels along the Ramapo River sold by the Indians on October 10, 1700 (Budke 1975a:77–78). Japough and Yawpaw were two of many spellings colonists also used to identify Tatapagh, an Esopus and Minisink leader who took part in land sales on both sides of the New York–New Jersey line between 1683 and 1715. His memory is commemorated on present-day maps as the name of Boy Scout Camp Yaw Paw, located along the Ramapo Mountains near Oakland.

Part 2

IMPORTS, INVENTIONS, INVOCATIONS, AND IMPOSTORS

ACAHELA (Monroe County, Pennsylvania). Acahela is the name of a camp and its approach via a road located where Tobyhanna Creek falls into the Lehigh River in Tobyhanna Township. In 1919 local Boy Scout council officials selected the name for their newly opened camp from what is remembered as an Indian word list identifying a suitable reflection of the locale's characters. Acahela as a Delaware expression meaning "at the joining of the waters," a suitable reflection of the locale's character. The word's similarity to Akela, one of the main characters in Rudyard Kipling's *Jungle Book* (1894), much beloved by Scouting's founders, probably helped secure its selection as the camp name, which remains on maps of the locale up to the present day.

ADIRONDACK. See **RONDACK**

ALLEGHENY (Northampton County, Pennsylvania). The name Allegheny most notably occurs in the Greater New York area as the name of a four-mile-long creek and an adjoining road located in the town of Upper Mount Bethel. Known as Hunters Creek as early as 1739, the stream has borne the name of the western Pennsylvania river and mountain range named for the legendary Allegewi Indians since at least 1805 (J. and L. Wright 1988:2). Spelled in a variety of ways, Allegheny is a popular street and business name that occurs widely in and around the metropolitan area and the greater Northeast.

ANAWANA (Sullivan County, New York). Today, Anawana Lake, Camp, and Road are located near Monticello in the Sullivan County town of Thompson. Similar-sounding Anawanda Lake and Anawanda Lake Road lie between Tennanah Lake and Callicoon Center at the western end of the county in the town of Fremont. Both Anawana and Anawanda are variant spellings of Annawan, the name of Wampanoag sachem Metacomet's war captain killed in combat with colonists in 1676 during King Philip's War in New England.

ANONA (Bergen County, New Jersey). Anona Lake is a small reservoir at the east end of Masonicus Road in the borough of Mahwah named for Anona, a fictional Arizona Indian maiden celebrated in a popular love song written in 1903.

APPALACHIAN. The U.S. Geological Survey includes all uplands in the Greater New York region within the Appalachian Highlands Physiographic Division's Atlantic Coast Physiographic Province, one of thirteen physiographic provinces that make up of the Appalachian Division, a chain of mountainous uplands that stretches across the eastern United States and Canada from Newfoundland to northern Alabama. The westernmost parts of the metropolitan area in New York and Pennsylvania are also included in what is referred to as the Appalachia cultural province by the Appalachian Regional Commission, established by Congress in 1965. The commission coordinates government programs in a thirteen-state area between southern New York and northern Mississippi.

Metropolitan area residents probably best know the name as the Appalachian Trail, the 2,132-mile federally managed hiking path that passes through the region on its way from Maine to Georgia. The name comes from Apalachen, an Indian village near present-day Tallahassee, Florida, first seen by Spanish explorers during the early 1500s. The name is thought to mean either "other side of the river" in the Apalachee language or "dwelling on one side," in the Muskhogean Hitichi tongue (in Bright 2004:44).

AQUETONG (Bucks County, Pennsylvania). Aquetong is the name of an eleven-mile-long creek, a spring, a lake, and a hamlet in Solebury Township. William Davis (1876:295) made the undocumented claim that Acquetong Spring was the Indian name of the Great Spring located on land in Solebury purchased by William Penn's secretary James Logan in 1702. MacReynolds (1976:12–14) suggested that locals calling it both Aquetong and Logan Spring took to referring to the place as Ingham Spring after New England immigrant Jonathan Ingham bought land there in 1747. Residents of the nearby hamlet of Paxon's Corners named their new post office Aquetong when it opened in 1884. Aquetong Lake was created by damming part of the creek in 1900. The name is also preserved by the Upper Aquetong Valley Historic District, listed in the National Register of Historic Places in 1987.

ASHROE (Sussex County, New Jersey). Lake Ashroe is an artificial pond located in the Kittatiny Mountain Boy Scout Reservation near Branchville. Often regarded as a word of Indian origin, Ashroe is the name of a village in County Limerick in Ireland.

ASKOTI (Orange County, New York). The Civilian Conservation Corps created Askoti Lake in Harriman State Park by damming an unnamed stream flowing into Lake Kanawauke in 1935. The name comes from a Mohawk word meaning "one side."

ASSINIWIKHAM (Passaic County, New Jersey). The 1,100-foot-high Assiniwikham Mountain is located in the Norvin Green State Forest. It is a recent concoction made up of Delaware words for "stone" and "house."

BASHER (Bergen County, New Jersey; Orange and Sullivan counties, New York). Basher Kill is a creek whose waters rise in the Sullivan County town of Mamakating. The creek flows south past Wurtsboro into the 2,213-acre Bashakill Wildlife Management Area before joining the Neversink River at Cuddebackville in the Orange County town of Deerpark. The name is identified with Bachom, associated with land also called Minnesinck in the seventeenth-century Jansson-Visscher map series published between 1650 and 1777. It most recognizably first appears in colonial records as the starting point of the boundary line marking off the lands sold by the Indians in the June 11, 1703, Minisink Patent deed, which began "at a certain place commonly called by the Christians Old Bashes Land" (manuscript on file in the Orange County Historical Society). Ruttenber (1906a:229–30) suggested that the name was a local equivalent of "*Bashaba*, an Eastern-Algonquian term for 'sagamore of sagamores'" used in northern New England during the early seventeenth century. More highly educated colonists living a century later who may not have ever heard of New England Indian Bashabas would probably have known about the word's entirely coincidental cognates—Basha, the biblical name of the third king of the Northern Kingdom of Israel, and *pasha*, the Turkic word for "ruler" that began entering English around this time, often in the form of *bashaw*. The 0.8-mile-long Bashes Creek that flows into the Hackensack Meadowlands in Randolph Township, New Jersey, is apparently a wholly distinct name honoring a local family bearing the surname.

BIG INDIAN (Ulster County, New York). The 3,710-foot-high Big Indian Mountain and the 33,500-acre Big Indian Wilderness Area are only two of the more notable locales associated with this mythological figure in present-day Catskill State Park. Although particulars have differed over time, the fundamental core of the legend recounting the tragic tale of the doomed love affair between an extraordinarily tall Esopus chief called Big Indian and a local settler's daughter has remained unchanged for more than two centuries.

BUCKWAMPUM (Bucks County, Pennsylvania). Buckwampum Hill and Road in Springfield Township is often regarded as a play word mingling the Indian word wampum with an English slang term for Indian man, a more proper word for a male deer, or the even more upstanding name of Bucks (from Buckinghamshire) County. The first part of the name instead more specifically memorializes Nicholas Buck, a German immigrant who settled in the area during the mid-1700s (W. Davis 1876:570).

CADJAW (Wayne County, Pennsylvania). Although the namesake of Cadjaw Pond and Cadjaw Pond Road is often thought to be an Indian, the pond built on the border of the borough of Honesdale and Cherry Ridge Township during the nineteenth century was most likely named for a resident bearing the Cornish surname also often spelled Cadjew and Cadjen.

CADOSIA (Delaware County, New York). Cadosia is the name of a six-mile-long creek that flows past a hamlet named after the stream as it makes its way to its junction with the East Branch of the Delaware River three miles northeast of the village and town of Hancock. First appearing on local maps during the early 1800s, the name probably comes from Edward Gibbon's popular account, in his multivolume *Decline and Fall of the Roman Empire*, published between 1772 and 1789 (e.g., Gibbon 1851:938), of the battle of Cadesia, fought in 626 at Al Qadisiyah in present-day Iraq. The Muslims' victory allowed them to seize "the wealthy province of Irak, or Assyria," from the defeated Persians. The Cadosia Valley post office took on the name of the creek when it opened sometime between 1850 and 1860 (Kaiser 1965).

CANADENSIS (Monroe County, Pennsylvania). The name of the Pocono resort community of Canadensis, "from Canada," is a Latinized variant of *kanata*, an Iroquoian word for home, house, or settlement. It primarily occurs as the second part of binomial (two-part) scientific names such as *Tsuga canadensis* (Northern North American hemlock; *Tsuga caroliniana*, "from Carolina," refers to the more southeasterly species of the tree), the namesake of the village set up at the locale during the mid-1800s by leather tanners using bark from locally logged hemlock trees to prepare hides.

CANOPUS (Putnam County, New York). Horton's Pond and Horton Creek were both given the name of this mythological local Indian sachem when the Clarence Fahnestock Memorial State Park was built in 1929. Canope, it turns out, was an actual Indian person, one of the few who tried to return to his home in the Upper Delaware Valley after the end of the Revolutionary War. Like several of his countryfolk who also tried to do the same thing, Canope was murdered by a local settler near his home around 1789 (Goodrich 1880:158, 221). The spelling of the name most closely resembles the Canopus Branch of the River Nile in lower Egypt.

CASTLE HILL (Bronx County, New York). Robert Bolton (1881[2]:264) probably started the local and wholly undocumented tradition identifying Castle Hill Neck as the site of an Indian fort allegedly seen by Dutch voyager Adriaen Block in 1614 or 1615. The locale first appeared in colonial records as Cromwells Neck, named for the family that settled there during the mid-1680s. Wealthy landowner and politician Gouverneur Morris Wilkins probably gave the name Castle Hill to the estate he built on the neck around the time he was a delegate to the New York revolutionary provincial congress in 1775. The name hearkened back to one of the several grand Castle Hill estates in Great Britain. Developers exploited the name's cachet to sell subdivided plots on the neck during the late nineteenth century. By 1901 the neighborhood grew large enough to support a public school. That same year Castle Hill's main thoroughfare, Avenue C, was changed to its present name, Castle Hill Avenue.

CAUDATOWA (Fairfield County, Connecticut). Caudatowa Drive is in the town of Ridgefield. Local residents familiar with the writings of avocational historians such as Daniel Teller (1878:12) have regarded Caudatowa as the original, albeit undocumented, Indian name for the Ridgefield area.

CHIKAHOKI (Passaic County, New Jersey). Heckewelder noted that Delawares of his acquaintance translated the name of the place they identified as *tschichohacki* at the present-day site of the city of Burlington, New Jersey, as "ancient cultivated land, or oldest planted ground," in reference to a tradition regarding the place as their earliest settlement on the Delaware River. The name occurs on present-day maps as Chikahoki Falls in the Norvin Green State Forest. It is a recent import from the Delaware River Valley, where it is perhaps best known in the form t'Schichte Wacki, a place name entered on the Jansson-Visscher maps (Campbell 1965) at the present-day Minisink Island locale.

CHILOWAY (Delaware County, New York). Chiloway is a hamlet located on the banks of the Beaver Kill in the town of Hancock. The community was named for Job Chilloway, a Unami Delaware Indian translator and frontier diplomat who, as far as the historic record indicates, never set foot in the village named for him in New York's Delaware County. Born in southern New Jersey, Job Chiloway was converted to Christianity by the Moravian brethren at Wyalusing, Pennsylvania, in 1770, and died in Ohio in 1791.

CHINTEWINK (Warren County, New Jersey). Whritenour thinks Chintewink sounds like a Delaware word, *tschinktewink,* "on the south or sunny side of the mountains." Chintewink Alley is the only place in the freestanding town of Phillipsburg that still retains the name on local maps. Uncorroborated local tradition holds that Chintewink was the name of a Lenape Indian village said to have been entered onto a version of the Jansson-Visscher map printed in Adriaen van der Donck's *Description of New Netherland* published in 1654.

CHIPPEWALLA (Litchfield County, Connecticut, and Dutchess County, New York). Chippewalla Road is located within the bounds of Macedonia Brook State Park in the Connecticut town of Kent. Slightly differently spelled North Chippawalla Road is situated just across the state line about seven miles to the southwest of Kent in the village of Wingdale, New York. James S. Muncey (who traces his ancestry to Maritime Province Canadian non-Indian forebears [personal communication with the author, December 6, 2011]) adopted the existing name of the North Chippawalla Road for his Chippawalla Properties, Inc., subdivision established in 2008. References to the Chippewalla Valley in the Amenia, New York, vicinity first appeared in local newspapers during the 1890s. The name was mentioned shortly thereafter, in 1910, as a variety of nut and as the name of Chippewalla Advance, a Holstein-Friesian cow listed in a cattle registry published in 1918.

CHODIKEE (Ulster County, New York). Chodikee is the name of a lake, a road, a hotel, and several other places around the lake in the town of Highland. The lake was created sometime during the late 1800s when local residents dammed a swampy section of Black Creek. Originally called Black Pond, the lake was given its current name about the time local businesspeople opened the Chodikee Lake Hotel in 1910. Two years later, Gustav Stickley, furniture maker, leading light of the American Arts and Crafts movement, and educational reform activist, established his Craftsmans Farms School for Citizenship on the lake alongside the resort. Stickley's school attracted fellow reformer Raymond C. Riordon to the locale in 1914. His Raymond C.

Riordon School, established on the site of the Craftsmans Farms institution after it closed, employed a curriculum stressing appreciation of Indian and other native values grounding students to "roots in American soil." Closing shortly after Riordan's death in 1940, the tract was acquired by the state of New York, which operated the Highland Training School for Delinquents there between 1957 and 1976. The former training school facility is presently the site of a private summer camp.

Area tradition holds that Chodikee (locally pronounced "shakatee") is an Indian word referring to signal fires. The name more properly may be traced to a different lake along another waterway bearing the name black: Lake Tchulkade along the Black River in Alaska. Tchulkade comes from the Athabaskan word *chuu kaadlai*, "water is there" (in Bright 2004:484).

CIRCLEVILLE (Orange County, New York). Impressed by Indian earthworks she saw while visiting Circleville, Ohio, local resident Mary Bull convinced her neighbors to adopt the name for their community in the town of Wallkill in 1841 (Vasilev 2004:45).

COHASSET (Orange County, New York). Cohasset Lake in Harriman State Park was built in 1924 and named by park director William A. Welch for the scenic village of Cohasset south of Boston on Massachusetts Bay. The name is said to come from a Massachusett word meaning "small pine place" (in Bright 2004:115).

COMMUNIPAW (Hudson County, New Jersey). Sometimes thought to be a Delaware place name meaning "landing place," the present-day Communipaw neighborhood in Jersey City instead more probably bears a rendering of the Dutch name of the community established on land at the locale called Pavonia, first purchased by Michiel Paauw in 1630.

CONASHAUGH (Pike County, Pennsylvania). Three-mile-long Conashaugh Creek flows from the town of Dingman into Delaware Township, where its falls into the Delaware River across from Namanock Island. Today, the creek is celebrated as the site of the Battle of Conashaugh, during which a number of local militiamen were killed and wounded in a fight with a mixed Tory and Indian raiding column from Niagara on April 20, 1780. Contemporary records noting that the battle took place three miles south of Milford, however, do not mention its location by name. The present-day name of the creek may come from the Conasauga region on the Georgia-Tennessee border. Union Army deserters moving into the remote creek valley evidently brought the name with them to the Delaware Valley. If this is the case, the name Conashaugh may come from a Cherokee word, *kanesega*, "grass" (in Bright 2004:117).

CONESTOGA (Bergen County, New Jersey, and Northampton County, Pennsylvania). Most often associated with the Conestoga wagon prairie schooners of wagon-train fame, Conestoga was a name of the Iroquoian-speaking Susquehannock Nation, whose homeland centered around the lower reaches of the Susquehanna Valley during much of the colonial era. The name appears as an import in Greater New York given to a street in Bethlehem, Pennsylvania, and one of several roads bearing Indian names in a Franklin Lakes, New Jersey, development. The name is traditionally translated as a Susquehannock word meaning "place of the immersed pole or cabin pole."

COS COB (Fairfield County, Connecticut). The name first appeared as Coscob Neck in an Indian deed to land in the area dated June 9, 1701 (Firestone Library, Princeton University, Scheide Collection, Indian Deeds:20). Present-day Greenwich Creek has gone by the names Cos Cob Creek and Asamuck River at various times in its history. Long thought to be an Indian name, Cos Cob is more likely a confection joining the name of the Coe family, one of whose members bought the neck in 1701, to cob, an old English word for seawall still in use when the purchase was made.

CRONOMER (Orange County, New York). Today, Cronomer is the name of a hill (elevation 725 feet), a county park, two neighborhoods (Cronomer Valley and Cronomer Heights), and several other places in and around the city of Newburgh. No known record of this name appears to predate the 1850s. Although local traditions associate the name with a benevolent local Indian chief, Ruttenber (1906a:130) thought that it was probably a folk Dutch rendering of a local settler's name.

COUNCIL ROCK (Nassau County, New York). Council Rock is a glacial erratic boulder located on Lake Avenue by the Mill Pond just west of the Oyster Bay village center. A state historical marker located near the rock states that it was the council place and center of the Matinecock Nation and the place where Quaker George Fox preached while visiting Oyster Bay in 1672.

CUPSAW (Passaic County, New Jersey). Today, Cupsaw Brook is a four-mile-long stream that rises at Shepard Lake astride the New York–New Jersey border. From there it flows south into Cupsaw Lake to the place where it falls into the Wanaque Reservoir one mile beyond the outlet of the Cupsaw Lake Dam. The place name first appeared in early-twentieth-century geological reports as a stream in the Ringwood Township mining district. Cupsaw Lake was built in 1932 by a private corporation as the centerpiece of a private resort and residential community maintained today by the Cupsaw Lake Improvement Association. Although Becker (1964:16) thought Cupsaw might be an Indian word (Whritenour suggests a possible Munsee etymology, *kphasuw*, "it is dammed up"), Vermeule (1925:253) made a persuasive case for a Dutch origin, suggesting "cooper's staves saw" from *kuip*, "tub or cooper's staves," and *zaag*, "saw."

DELAWANNA (Passaic and Warren counties, New Jersey). This toponym combines contractions of the first two names of the Delaware, Lackawanna, and Western Railroad. Today, the name Delawanna appears at opposite ends of the Boonton Branch of the line that ran along much of the Morris Canal's route between Montclair and Hackettstown. Delawanna Station, at the eastern end of the line, was built in 1925 in the city of Clifton, where Delawanna Avenue crosses the tracks of the present-day Montclair-Boonton New Jersey Transit line, first built in 1869. Delawanna Creek, at its western end, is a 4.5-mile-long stream that flows through Knowlton Township from its headwaters near the village of Knowlton into Delaware Lake and on to its confluence with the Delaware River at Ramseyburg.

ESPANONG (Morris County, New Jersey). Espanong is the name of a road and a hamlet on the shore of Lake Hopatcong in Jefferson Township. The name appears to be have been invented sometime around the turn of the twentieth century by the proprietor of the Espanong House resort hotel. It seems to have been constructed by replacing the *ing-en* ending of the then-familiar Jersey City Indian place name variously spelled Hespatingh (in 1657), Espetingh (in 1671), Espating (in 1674), and, most recognizably, Espaten (in 1668) with the local Indian place name suffix *ong* (as in Hopatcong and Netcong). Whritenour is not alone in thinking that Espanong sounds almost exactly like a Munsee word, *eespanung*, "place of raccoons." Local residents evidently using published Delaware dictionaries available in local libraries made the same finding (e.g., Hutchinson 1945:13). Noting that Espanong and nearby Raccoon Island mean the same thing in different languages, Becker (1964:18) intimated that Espanong was the inspiration for the latter name. The opposite may instead be the case. Raccoon Island was an upscale resort destination that attracted a well-to-do clientele during the early 1900s. This may have led the owner of Espanong House to adapt and adopt its more romantic-sounding Indian nomenclatural doppelganger in hopes that some of the glamour of Raccoon Island would rub off on his more modest establishment.

GANAHGOTE (Ulster County, New York). The name Ganahgote was given to the postal village called Tuthilltown at the place where the Shawangunk Kill flows into the Walkill River, just west of the village of Gardiner, around the turn of the last century. It was evidently thought to be a slightly respelled version of *Ganasote,* an Iroquois word for longhouse.

GLEN ONOKO (Carbon County, Pennsylvania). Lehigh Valley Railroad officials gave the name Glen Onoko to what they called a "rustic depot" built at Moore's Ravine in 1873 (Nigro 2002:228). Onoko was the name of another legendary Indian princess condemned to suffer an appropriately tragic literary end in the pages of a breathless Victorian romantic novel.

HACKLEBARNEY (Morris County, New Jersey; Orange County, New York; Carbon County, Pennsylvania). Hacklebarney State Park lands currently straddle the border between Washington and Chester townships in the heart of the old North Jersey iron-mining district. Several writers have thought that Hacklebarney was an Indian place name. Others have suggested that it originally was the name of a local mine foreman named Barney Hackle. And a few have insisted it was actually his nickname, Heckle Barney. It is more likely that the name comes from the village of Hacklebarney in County Cork, Ireland, a place surely well known to many of the ironworkers who worked at the Hacklebarney Mine and other pits in the region. Differently spelled versions of the name adorn such other locales in Greater New York as the hamlet of Hacklebernie at the mouth of Mauch Chunk Creek in Carbon County, Pennsylvania, and High Barney Road in the city of Middletown, New York.

HASECO (Westchester and Queens counties, New York). Haseco Avenue in the city of Portchester is sometimes thought to be an old Indian name. I have not been alone in assuming that its lookalike, Hassokie Creek, a stream that flows into Jamaica Bay, shared a similar pedigree (Grumet 1981:13). William Tooker (1911:70, 102), for example, was only one of the first to suggest such a connection. It turns out, however, that Haseco and Hassokie both come from the old English word "hassock" that referred to a dense clump of grass.

HAWTREE (Queens County, New York). The name Hawtree currently graces Hawtree Basin and the Hawtree Basin Pedestrian Bridge in Queens County, New York. Although it sounds Indian and appeared in local colonial documents as early as March 15, 1656 (Grumet 1981:13), it is an English word referring to hawthorn trees.

HEAPTAUQUA (Westchester County, New York). Heaptauqua Lake in the town of New Castle is a joke name combining "heap talking" with the Indian-sounding *qua* ending from nearby Chappaqua. The name was invented by local water company owner Victor Guinzburg in 1902.

HIAWATHA (Morris County, New Jersey, and Wayne County, Pennsylvania). Places called Hiawatha village and Lake Hiawatha are located in Scott Township, Pennsylvania and Parsippany–Troy Hills Township, in New Jersey. Both paired sets of place names were drawn directly from the fictional main character of Longfellow's poem *Song of Hiawatha* (1855), given the same name as the traditional founder of the Iroquois Confederacy.

HOLICONG (Bucks County, Pennsylvania). Holicong is the name of a hamlet, a road, a municipal park, and several other places in Buckingham Township. The present-day hamlet of Holicong was called Grinville during the early nineteenth century. Postal authorities changed its name to the more dignified Greenville when they opened the first post office at the locale in 1881. Local residents subsequently adopted a respelled version of the name of a spring they called Hollekonk Well for the name of their hamlet and post office (MacReynolds 1976:195–97). Holicong Village Historic District was listed in the National Register of Historic Places in 1980.

HOMOWACK (Sullivan and Ulster counties, New York). Folklorist Charles Gilbert Hine (1908:101) held that Homowack was an Indian word for "the water runs out." Other writers have suggested different Delaware and Iroquois etymologies. Present-day Homowack Kill is a small stream that rises on the western slopes of the Shawangunk Ridge at the northeastern corner of the town of Mamakating in Sullivan County. It flows north into the Sandburg Creek at Spring Glen in the Ulster County town of Wawarsing. Locals knew the lower course of Sandburg Creek below its confluence with Homowack Kill as the Leuren Kill (also called Lunankill and noted as the Lurenkill in Spafford [1813:109]) at the time the Delaware and Hudson Canal was built through the area in 1831. The canal town on the creek's banks was initially called Red Bank. The creek, the village, and its church were renamed Homowack by 1843. The post office opened in the hamlet in 1852 was also given the name. In 1863 local businessmen incorporated a firm christened the Homowack and Fallsburgh Plank Road Company in order to build a road for wagons carrying hemlock bark from nearby hills west of Homowack and supplies for the Metropolitan Mine workings in the Shawangunks just east of the town. A portion of the original route survives today as Plank Road. Homowack villagers changed the place's name to Spring Glen in 1891 in hopes of attracting summer tourists traveling to the Catskills on the New York, Ontario, and Western Railroad built through the area during the 1880s. Tourist capacity and appeal increased after village buildings emptied by the closing of the Delaware and Hudson Canal in 1898 were converted into boardinghouses. Several of these moved into more spacious accommodations as the area's popularity increased. The best known of this second generation of resorts was the sleek, modern Homowack Lodge and Golf Course built during the late 1950s. It was a popular tourist destination whose heyday ended by the time the Catskill resort industry collapsed during the 1970s. Subsequent attempts to revive Homowack and other resorts in the area have not encountered much success to date.

INDIAN is an extremely widespread generic word used to establish some sort of connection between a locale and the region's first people. A partial list of such generics not connected in any known way to actual Indian people or places in Greater New York includes:

Cold Indian Springs (Monmouth County, New Jersey) is the name of a street and an unincorporated community in the city of Asbury Park.

Indian Field (Bergen County, New Jersey) is one of several locales bearing this name in the region.

Indian Head (Morris and Warren counties, New Jersey; and Greene and Ulster counties, New York). Indian Head Cliff is located at the Delaware Water Gap in Knowlton Township. Indianhead Road is in the city of Morristown. The 3,753-foot-high Indian Head Mountain and 35,000-acre Indian Head Wilderness Area are located in Catskill State Park.

Indian Hill (Bergen and Monmouth counties, New Jersey; Orange County, New York; Carbon County, Pennsylvania). Indian Hill is the name of hamlets in Southfields, New York, and Weissport, Pennsylvania. It is also a school name in the New Jersey communities of Holmdel and Oakland.

Indian Lake (Morris County, New Jersey). The reservoir and residential community of Indian Lake in Denville Township, first constructed as Lenape Lake around 1920 and given its current name in 1923, is only one of many places given this name in the region.

Indian Mountain Lake (Carbon and Monroe counties, Pennsylvania) is the name of a lake, a dam, and a residential community located along the Monroe-Carbon county line just northwest of Mount Pohopoco.

Indian Orchard (Wayne County, Pennsylvania). The name of the hamlet of Indian Orchard is an import from New England. The original name comes from a colonial town in Massachusetts that had been the home of some of the first New England settlers who came to northeastern Pennsylvania after 1761.

Indian Park (Orange County, New York) is an unincorporated community on the shores of Greenwood Lake.

Indian Point (Westchester County, New York, and Pike County, Pennsylvania). The Indian Point Energy Center in Buchanan was built during the early 1960s at the former site of the Indian Point Amusement Park. The name also adorns an unincorporated community in Dingman Township, Pennsylvania.

Indian Pond (Bronx County, New York). Two examples of the many Indian Ponds in the region are located in Bronx borough's Crotona Park and the Fieldston neighborhood in Riverdale.

Indian Ridge Road (Orange County, New York) is located in the hamlet of Westtown.

Indian Road (New York County, New York) and the Indian Road Playground in the Inwood section in northern Manhattan are both named for nearby archaeological rock shelter sites excavated during the first decade of the twentieth century. A long-fallen elm in the park was thought to have shaded Indians and Dutchmen while they dickered over the price for Manhattan in 1626. The area was also the site of an "Indian Life Village" managed by people tracing descent to Indian ancestors during the 1910s and 1920s. The place was later the locale of powwows held by American Indian people and Indian hobbyists living in the metropolitan area still remembered by older city residents.

Indian Rock (Passaic and Somerset counties, New Jersey; Ulster County, New York). Indian Rock is another frequently encountered place name in the region. Examples of places in New York bearing the name include Indian Rock Cliff in the town of Esopus and the hamlet of Indian Rock in the town of Wawarsing. The name also appears in New Jersey as Indian Rock Road in Warren Township and Indian Rock Trail in the Ramapo Mountain State Forest.

Indian Run (Somerset County, New Jersey) in Whitehouse Station is only one of many subdivisions given the name in the region.

Indian Swamp (Pike County, Pennsylvania) in Greene Township is one of a number of similarly named places in the metropolitan area.

IOSCO (Passaic County, New Jersey). Developers creating Lake Iosco in Bloomingdale Township in 1925 gave the lake and the subdivision they built alongside it the name Iosco, which Henry Rowe Schoolcraft translated as an Ojibwa name meaning "water of light."

JAMAICA (Queens County, New York). For nearly two hundred years local tradition has held that the etymology of the Queens County place name Jamaica can be traced to a Delaware word that Tooker (1911:75) identified as *tamaqua* or *tamaque*, "beaver." The name was thought to apply to both the place and the Indian tribe or band that lived there. Although I could find no evidence of an Indian community identified by the name, I originally supported the beaver etymology (Grumet 1981:16). Since then, no further evidence supporting a Delaware origin has turned up for the name Jamaica, which presently adorns a neighborhood, a boulevard, several lesser roads, and much else in the old Queens County seat. The only colonial document mentioning beaver and Jamaica together is the November 25, 1656, purchase confirmation for English settlers "living at the new plantation near unto the Bever Pond commonly called Jemaico" (O'Callaghan and Fernow 1853–87[14]:504–505). Neither this nor any other document from the era states that Jamaica was an Indian name, that it was a word for beaver, or that Jamaica was the Beaver Pond. When read as something written by residents of a place that the Dutch formally called Canarise, the passage seems to refer to the new plantation evidently commonly called Jemaico near the Beaver Pond. The September 13, 1655, English deed to land in the area is sometimes cited to confirm the connection linking a beaver pond and Jamaica. Examination of original copies of the deed bearing that date disclose no mention of either place name (New York Public Library, Papers Relating to the Bounds of Jamaica, 1655–85 [39]:M105). Other documents show that Canarise was subsequently renamed Rustdorp, Dutch for "restful place," in 1660. Soon afterward, the place was renamed the town of Crawford by the English who seized New Netherland from the Dutch in 1664. Many documents written during this period show that local English settlers continued to informally call the place Jamaica. Provincial officials finally accepted Jamaica as the town's formal name by the time Dutch forces that had retaken the province in 1672 returned it to English rule a year later. It seems likely that the name Jamaica came to Queens in much the same way it made its way to Boston and London, as a name proudly proclaiming connections to the commercial crown jewel of the British West Indies whose name meant "land of wood and water" in the Jamaican Arawack language.

JENNY JUMP (Warren County, New Jersey). The 4,200-acre Jenny Jump State Forest is managed by the New Jersey Division of Parks and Forestry in Hope Township. The name commemorates a legendary leap said to have been made to avoid the unwanted attentions of pursuing Indians.

JIM THORPE. See **MAUCH CHUNK**

KAHAGON (Orange County, New York, and Lackawanna County, Pennsylvania). Kahagon appears in two places in the Greater New York Area. Camp Kahagon in Harriman State Park is operated by the YMCA of Bergen County. Farther west, Lake Kahagon is located in the borough of Moscow in Pennsylvania. Although Kahagon is sometimes thought to be a Delaware place name, it is instead an Iroquois word meaning "in the forest."

KANAWAUKE (Orange County, New York). Kanawauke Lake in Harriman State Park was created after a dam holding back the waters of Little Long Pond was enlarged in 1918. Park general manager and chief engineer William A. Welch christened the lake *kahnawà:ke*, "place of the rapids" (in Bright 2004:200), after the well-known Mohawk community in upstate New York that ultimately reconstituted itself on the outskirts of Montreal, Quebec, during the late 1600s. Kanawauke Lake was the site of New York City Boy Scout summer camps (spelled Kanowahke) from 1918 until 1930.

KANOUSE (Passaic County, New Jersey). The 1,190-foot-high Kanouse Mountain is located just southeast of the hamlet of Newfoundland in West Milford Township. Although often thought of as an Indian name, it is actually the surname of a family of German origin.

KAUNEONGA (Sullivan County, New York). Kauneonga is the name of a lake and a hamlet in the town of Bethel. Catskills poet Alfred B. Street (1845:75) invented Kauneonga as an Indian name for White Lake whose purported translation, "two wings," metaphorically represented the way the lake's two connected parts looked on maps. Local residents gave Street's name to the northern wing of the lake sometime during the early 1900s. The hamlet and its post office at its northeastern end both later adopted the name Kauneonga Lake.

KEMAH (Sussex County, New Jersey). Lake Kemah is a summer resort built in Hampton Township during the late 1920s, and the name of the road that connects it to the Morris Turnpike. The name was borrowed from the popular Gulf Coast Kemah tourist destination near Galveston, Texas.

KENOSIA. See **KENOZA**

KENOZA (Fairfield County, Connecticut; Sullivan and Ulster counties, New York). Variants of Kenoza, an Ojibwa word for "northern pike" that entered English as *Keno'zha*, "the pickerel," through Longfellow's poem *Song of*

Hiawatha (1855), occur in several places located in and around Greater New York. These include the New York locales of Kenoza Lake in Sullivan County, Kenozia Park and Pond in the town of Hurley in Ulster County, and Kenosia Lake (originally Mill Plain Pond) in the city of Danbury, Connecticut (Bailey 1896:5–7). Well-known instances of the name elsewhere include the city of Kenosha, Wisconsin, and Kenoza Pond in Massachusetts, named by poet John Greenleaf Whittier.

KEOWA (Sullivan County, New York). Ten Mile River Scout Reservation's Camp Keowa is a slightly respelled iteration of Kiowa, the name of a notable Plains Indian nation.

KIAH (Fairfield County, Connecticut, and Putnam County, New York). The Kiah portion of the name of the forty-three-acre Kiah's Brook Refuge/Titicus Preserve in the town of Ridgefield is often regarded as an Indian word. It was more probably named for local settler Hezekiah Scott (Sanders 2009). The 1,038-foot-high Kiah Hill in Clarence Fahnestock Memorial State Park evidently bears the shortened nickname of another settler named Hezekiah.

KIAMESHA (Sullivan County, New York). Kiamesha Lake and the Kiamesha Lake community are located just north of the city of Monticello in the town of Thompson. The name comes from Kiamichi, the river that flows through Ozarks Mountain resort communities in Arkansas (in Bright 2004:216). Kiamichi comes from the Caddo language and has been widely associated with mountain resorts in many places across the United States since the late nineteenth century. Present-day Kiamesha Lake was created when a dam was built at its outlet sometime before 1799. Known early on as Pleasant Lake, it was renamed Kiamesha Lake during the 1890s. The place soon became a major center of the Catskill Mountains summer resort industry. Its largest facility, the Concord Hotel, operated at Kiamesha Lake between 1935 and 1998. The defunct resort's buildings were completely demolished in 2008. Today, Kiamesha Lake is a mixed residential and resort community.

KISSENA (Queens County, New York). Kissena Park and the Kissena Park neighborhood are located in the borough of Queens within the borders of the old Long Island town of Flushing. Samuel Bowne Parsons imported Kissena, an Ojibwa word meaning "it is cold," as the name for the nursery he operated at the locale between 1839 and 1906. Japanese maples, white mulberries, weeping beech, and several popular species of magnolias are among the many plants introduced into the United States by Kissena Nursery staff. Having acquired nearby Kissena Lake in 1904, the New York City Parks Department completed its plan for its Kissena Park unit by purchasing the nursery shortly after Parsons's death in 1906. A fourteen-acre fragment of the original nursery grounds rediscovered in 1981 is today preserved in the park as the Kissena Park Historic Grove. The name also adorns nearby Kissena Boulevard and a number of other places in and around the Kissena Park neighborhood.

KOHANZA (Fairfield County, Connecticut). Upper and Lower Kohanza lakes are located close to one another on Kohanza Brook in the city of Danbury. Its first appearance as the "land at Cohansey" in a document dated 1745 suggests a connection of some sort with the identically spelled Indian place name from the southernmost part of New Jersey. Whritenour thinks Cohansey in New Jersey sounds like a Southern Unami cognate of a Northern Unami word, *gahansik*, "that which is taken out." The name quickly morphed into Conhansa in 1767, Cohanzy Pasture in 1776, Cohanzy Orchard in 1780, and Cowshandy Lot in 1798 (Bailey 1896:9). People in the area subsequently began using a version of the name spelled Kohanza, an English and Welsh family name, when referring to the brook and the reservoir built along its course in 1860. In 1869 waters released by a catastrophic breach in the Kohanza Reservoir Dam killed eleven local residents. Today, smaller dams hold back the now separately impounded waters of Upper and Lower Kohanza lakes.

KOWAWESE (Orange County, New York). Kowawese State Unique Area is a state-owned 102-acre park managed by the Orange County New York Department of Parks, Recreation, and Conservation at Plum Point in New Windsor. The name comes from the Narragansett word *kowawese*, "young or small pine," that Ruttenber (1906a:90) suggested was a cognate for Gowanus in his book of Hudson River Valley Indian place names.

LACKAWANNA (Warren County, New Jersey; Lackawanna and Luzerne counties, Pennsylvania). Heckewelder thought Lackawanna reminded him of the Delaware words *lechawahhannek*, "forks of the river," and *lechauhanne*, "forks of a river." Nora Thompson Dean traced the origin of the name to a Southern Unami word, *lèkaohane*, "sandy creek or river." Lackawanna appears in the metropolitan area as an import from the Lackawanna River Valley that flows through the city of Scranton into the Susquehanna River at the Wyoming Valley. It occurs widely in Greater New York as a name associated with the valley's anthracite coal fields and the Delaware, Lackawanna, and Western Railroad, which carried coal from the fields across the region to ferry slips along the banks of the Hudson River. Lake Lackawanna was originally gouged out of the ice-age gravels located above the village of Stanhope, New Jersey, that provided the spoil rubble used to raise up the adjacent Lackawanna Cut-Off High Line in 1909. Portions of Lackawanna County spill over the mountain ridges separating the Susquehanna drainage from Greater New York's Upper Delaware Valley. The Pennsylvania Department of Conservation and Natural Resources manages units of the 27,345-acre Lackawanna State Forest at the westernmost edges of the Lehigh Valley, acquired by the state to prevent their destruction during the late 1890s.

LAURENCE (Middlesex County, New Jersey). The Laurence Harbor locale in Sayreville Township was first identified as Arromsinck in December 12, 1663 (O'Callaghan and Fernow 1853–87 [13]:316–17). Whritenour thinks Arromsinck sounds like a Delaware word, *allumesink*, "place of small dogs." Similar-looking Arowonenoc was subsequently identified as a neck of land in a deed to land in the area dated May 22, 1676 (Monmouth County Records, Deed Book B:11–14).Variant spellings such as Arawense suggest a possible translation of *alluns*, "arrow" (Whritenour in Boyd 2005:449). Conflation of Arawense with the first name of Laurence Lamb, builder of the country club he named after himself at Laurence Harbor, resembles the way Bedouins pronounced T. E. Lawrence's last name as "Ourans" in the movie *Lawrence of Arabia*. Boyd (2005:449) thought that such a resemblance was accidental and played little if any part in the naming of the present-day locale.

LENAPE (Union and Sussex Counties, New Jersey). Whritenour thinks Lenape sounds almost exactly like a Munsee word, *lunaapeew*, "Indian or Delaware person." The name Lenape appeared in Greater New York only once as Ninnepauues, the Mahican form of the word used to identify the mostly Munsee Delaware Indians who signed the May 6, 1755, deed to land in northeastern Pennsylvania (Boyd and Taylor 1930–71[1]:308–14). Despite this, Lenape has long been a very popular camp, street, park, and institutional name throughout the region. Lake Lenape in Andover Township, Sussex County, and the Union County Park Commission's sprawling Lenape Park built by 1930 are only two of the many places presently bearing the name in the metropolitan area.

LITTLE SILVER (Monmouth County, New Jersey). The borough of Little Silver is located in Shrewsbury Township. The local post office changed its name from Parkersville to Little Silver in 1879. Sometimes thought to be a rueful Indian reference to the insufficient payments they had to accept for their land, the locale is actually named for Little Silver, the original Devonshire home of the community's founding Parker family in England.

LOMMASON (Warren County, New Jersey). The name Lommason, attached to a scenic glen and the road that runs through it in White Township, is often thought to be an Indian name. It is actually the surname of a family in the area.

LONG HOUSE (Bergen County, New Jersey, and Orange County, New York). An English name for a possibly Indian place. Long House Creek is a stream in the Wallkill River drainage system. Beginning at its source in New Jersey just above Upper Greenwood Lake, it flows north to join Waywayanda Creek in the Orange County town of Warwick. Ruttenber (1906a:137–38) thought the name probably referred to an otherwise undocumented Indian longhouse located somewhere on or near the stream.

MALAPARDIS (Morris County, New Jersey). Hanover Township residents have long regarded the place name Malapardis given to a local pond, brook (also known as Stony Creek), park, and road as an Indian word. It more likely comes from an eighteenth-century English traveler's expression for an uncomfortably shared inn chamber said to mean "saucy companion," based on the Latin *mala*, "bad," and *pardis*, "partner."

MANATICUT (Passaic and Sussex counties, New Jersey). Manaticut Point and Trail are located in the Norvin Green State Forest. Manticut Road is in the Sussex County community of Highland Lakes in Vernon Township. Each is named after the Manaticut River in Massachusetts.

MANHASSET (Monmouth County, New Jersey, and Nassau County, New York). Manhasset, first documented as Menhansack and Manhansett (Tooker 1911:92–94), was originally the Indian name for Shelter Island at the eastern end of Long Island. Residents on Matinecock Neck adopted the name in the form of Manhasset Bay in 1837 as a replacement for Cow Harbor and the community at the base of Cow Neck, whose post office had been called Head of Cow since 1812 (Kaiser 1965). The name has since spread from Long Island to other locales such as Manhasset Creek, a small inlet in New Jersey that falls into the Shrewsbury River between the borough of Monmouth Beach and the city of Long Branch.

MANTOLOKING (Ocean County, New Jersey). The name currently graces Mantoloking Borough (established in 1911) and the Brick Township municipalities of Mantoloking Shores, South Mantoloking Beach, and West Mantoloking. The Mantoloking Bridge joins West Mantoloking on the mainland to its namesake communities on the narrow barrier beach separating Barnegat Bay from the Atlantic Ocean. The name was meant to identify the locality as the land of the Mantua Indians, regarded as the Lenni Lenape tribe that lived in the area. It was invented by Frederick W. Downer, a founder of Mantoloking who named the place in 1881.

MARATANZA (Ulster County, New York). Fried (2005:113) thinks Ruttenber's (1906a:145 nn.) translation of Maratanza as a word "from Old English *Mere*, 'a pond or pool,' and *Tanze*, 'sharp,' or offensive to the taste" is questionable. Today, Lake Maratanza is the highest of the hilltop lakes along the Shawangunk Ridge in the town of Wawarsing. Fried (2005:111–33) has shown that the lake currently bearing the name is not the original Lake Meritange mentioned as a major boundary marker fixing the location of the southwestern corner of the lands taken up by Governor Dongan's Indian deed dated September 10, 1684. Fried put so much effort into the search for the present-day location of the original Meritange that it seems almost churlish to suggest any locale other than the unnamed pond, dissected by Interstate 84 midway between Port Jervis and Middletown, New York, that he identified as the likeliest candidate for the locale. Crossing over the same land, I suspect that beaver dams blocking the stream flowing out of the presently artificially impounded mountaintop basin of Hawthorne Lake (formerly named Hathorn, the family name of the Orange County militia colonel who led the local volunteers defeated by the mixed force of Loyalist Indians and Tory Rangers led by Mohawk captain Joseph Brant at the Battle of Minisink Ford on July 22, 1779) could have created the "water pond lying upon the hills called Meratange" in 1684. Although the stream draining Hawthorne Lake flows west into the nearby Neversink River, away from the "river called Peakkaensinck" mentioned in the deed as flowing from Meratange, the headwaters of the Shawangunk Kill (known as Pakanasink Creek in colonial times) rise less than a mile from modern-day Hawthorne Lake along the mountain ridge at about the same altitude as the lake.

One place Fried has conclusively shown was not Meratange is present-day Lake Maratanza. He shows how Minisink patentees trying to increase the amount of land taken up by their purchase got Indians and colonists to identify the then-unnamed lake atop Sam's Point as the 1684 deed Meratange water pond.

Whatever its location, Maratanza is frequently regarded as an Indian name. Working from this assumption, the U.S. Navy has given this name to two vessels belonging to classes often bearing Indian names. The first was a wooden side-wheeler steamer (built in 1861) that saw service as a gunboat in the federal blockade of Southern ports during the Civil War. The second, a private vessel (built in 1941; original name S.S. *Virginia*), acquired by the U.S. Navy in 1942, was renamed U.S.S. *Maratanza* in 1944. This vessel served as a patrol craft operating along the Atlantic Coast defense perimeter during World War II.

MASKENOZHA (Pike County, Pennsylvania). The name of Maskenozha Lake in Lehman Township comes from *maashkinoozhe*, an Ojibwa word for "muskellunge" (Bright 2004:304). The word first entered Canadian French as *masquinonge* before being adopted into English as muskellunge.

MASSAWIPPA (Orange County, New York). Massawippa Lake in Harriman State Park was created by damming a wide stretch of meadow along Popolopen Brook in 1934. The name was imported into the region as a homage to the very popular fishing and camping resorts at Lake Massawippi and the Massawippi River in the Canadian province of Quebec.

MATAWA (Sullivan County, New York). Matawa Lake is a small pond built in 1949 along the Little Beaver Kill in Catskill State Park. A popular camp name widely used in the United States and Canada, Matawa is usually regarded as an Ojibwa word for "fork or confluence." The ten confederated Ojibwa and Cree communities constituting the Nishnawbe Aski Nation in Ontario use that translation of the word for their Matawa First Nations Tribal Council. Mattawa can also be translated as a Central Algonquian word for "enemy" (Bright 2004:273).

MAUCH CHUNK (Carbon County, Pennsylvania). Mauch Chunk Ridge, Lake, Road, Creek, and National Register Historic District (listed in 1977) are surviving examples of the Delaware Indian name long associated with the section of the upper Lehigh Valley that was once Pennsylvania's premier mining, transportation, and tourist center. The name itself is an authentic-enough Indian word. John Heckewelder said it was a Delaware word, *machtschúnk,* meaning "the bear's mountain." Lehigh Coal and Navigation Company founder Josiah Wright probably obtained the name from Heckewelder sometime between 1815 and 1818 as an appropriate Indian name for the prominent local feature called Bear Mountain just south of Bear Creek on the east bank of the Lehigh River. Wright subsequently extended the name to take in the ramshackle collection of huts then known as Coaltown located at the mountain's base.

Much happened at Mauch Chunk between its founding in 1818 and 1954, when borough officials changed the community's name to Jim Thorpe. Wright and his successor, Asa Packer, threw up the net of roads, canals, and rail tracks that carried anthracite from the coal fields on the other side of Mauch Chunk Mountain to the head house of the 8.7-mile-long Mauch Chunk Switchback Gravity Railroad (listed in the National Register of Historic Places in 1976). Carried down the railroad's steep incline, which later became a tourist thrill ride sometimes regarded as America's first roller coaster, the coal was then dumped into canal boats and railcars at loading platforms and chutes in the borough of Mauch Chunk for shipment to Philadelphia and New York. The place became the seat of Carbon County in 1843 and was the location of the widely covered Molly Maguires trial, in which four men suspected of being members of the Irish American secret society known by that name were found guilty of murder in 1876 and hanged together one year later.

The hard coal industry was still booming when owners of resorts built on hilltops with commanding views of the surrounding terrain promoted Mauch Chunk as "the Switzerland of America." Both industries had collapsed, however, by the time officials looking for ways to put their fast-decaying community back on the map renamed their borough Jim Thorpe in 1954 after persuading the recently deceased athlete's family to locate his tomb there. The mausoleum remains at the locale today, although family descendants have sued for the return of his remains in accordance with the provisions of the Native American Graves and Repatriation Act of 1990.

MAUWEEHOO (Fairfield County, Connecticut). Lake Mauweehoo and nearby Mauweehoo Hill in Sherman, Connecticut, bear the name of a prominent Schaghticoke Indian family.

MEENAGHA (Ulster County, New York). Mount Meenagha is a 1,750-foot-high Shawangunk Ridge mountain located in the village of Cragmoor in the town of Wawarsing. Ellenville businessman Uriah E. Terwilliger adopted the name from Longfellow's poem *Song of Hiawatha* for the resort he operated there from 1882 to 1922. The name in the poem, Meenh'ga, from the Ojibwa word *miinagaaswansh*, "blueberry," advertised one of the resort's, and the region's, premier summer attractions. The name also survives as a street name at the locale.

METAUQUE (Sullivan County, New York). Metauque Lake, near the village of Glen Spey in the town of Lumberland, was formed by a dam built across a brook flowing into the Mongaup River's Rio Reservoir one mile farther southwest. Metauque Lake has been on local maps since at least 1860 (French 1860:646). Efforts to track down the history of the name have thus far found only that Metauque sounds much like a Munsee word, *míhtukw*, "tree," listed in Delaware language dictionaries (e.g., O'Meara 1996).

MICHIKAMAU (Orange County, New York). The Greater Bergen County YMCA has operated Camp Michikamau in Harriman State Park since the 1980s. The name was drawn from a popular fishing and canoeing lake in Labrador.

MINEOLA (Nassau County, New York, and Monroe County, Pennsylvania). Most New Yorkers associate the name Mineola with the Long Island village in the town of North Hempstead. The place was given the name in 1858 and was incorporated as a village in 1906. Local traditions hold that Mineola is a shortened version of the name of a Delaware chief named Miniolagameka. Meniolagameka (Whritenour suggests it is a Munsee word for "oasis," earlier translated by Heckewelder as a Delaware word for a "rich or good spot within that which is bad or barren") was actually a mostly Munsee Delaware Indian town in a part of Pennsylvania's Lehigh Valley that the inhabitants were forced to abandon during the 1750s. The locale became a popular tourist destination during the 1830s. Lake Mineola (a glacial kettle hole) and the nearby hamlet of Neola in the Poconos still mark the places where Hempstead residents first discovered the name they later adopted for their community. Mineola is a Lakota word meaning "many waters." Its distribution is widespread; particularly well-known locales bearing the name are located in Florida and South Dakota.

MINNEWASKA (Ulster County, New York). Lake Minnewaska is a cliff-lined, glacially formed lake located at a high point of the Shawangunk Ridge. It lies within the 21,096-acre Minnewaska State Park Preserve managed by the New York State Office of Parks, Recreation, and Historic Preservation. Minnewaska is also a lake in Minnesota adorned by a Lakota name meaning "good water." The name arrived in New York when Alfred H. Smiley, who with his brother Alfred K. founded Mohonk Mountain House in 1869, bestowed it on the lake and the resort he opened on a cliff overlooking it in 1879. New York State bought the lake and the land around it in 1987.

The state opened its Minnewaska State Park Preserve in 1993, purchasing its most recent addition of 2,500 acres in 2006.

MOHANSIC (Westchester County, New York). The place name Mohansic is not known from colonial records. A railroad report published in 1867 noted that today's Mohansic Lake was then known as Keakles or Mohansic Pond. The Mohansic Golf Course, Mohansic Creek, and several streets also bear this place name in and around the town of Yorktown. Present-day Franklin D. Roosevelt State Park (opened in 1982) was first established as Mohansic State Park in 1922. The park is located on the grounds of the Mohansic State Hospital for the Insane and the New York State Training School for Boys. Although building began for these institutions in 1909, strong local opposition forced the closing of both facilities by 1918. Mohansic Lake is said to have been known as Crom Pond during the nineteenth century. Today, the Mohansic Creek runs from Mohansic Lake into present-day Crom Pond and on to its junction with the Muscoot River just below the Amawalk Dam.

MOHAWK (Monmouth and Sussex counties, New Jersey). Mohawk is a frequently encountered Indian place name in Greater New York. Unlike the Mohawk Branch of the Delaware River discussed earlier, all other occurrences of the name in the metropolitan area are transplants. Perhaps the most notable of these is Lake Mohawk, a privately owned artificial lake and residential community in Sparta and Byram townships in New Jersey. Developers Arthur D. Crane and Herbert L. Closs oversaw construction of the dam across the headwaters of the Wallkill River that flooded Brogden's Meadow whose impounded waters formed Lake Mohawk between 1926 and 1928. Although shops at the development's town center were built to resemble alpine chateaus, both entrepreneurs strongly encouraged Indian associations from the start. They gave the name Winona Parkway to the main development entrance road, operated an "Indian Village" complete with tepees and people in Indian costume during the 1940s and 1950s, and named the development center White Deer Plaza after one of the Indian Village's employees who traced her descent to Native ancestors. A smaller body of water in the region, Mohawk Pond, is located many miles southeast of Lake Mohawk at the head of Little Silver Creek in Shrewsbury Township.

MOHEGAN (Fairfield County, Connecticut, and Westchester County, New York). Often regarded, especially during the nineteenth century, as a synonym for Mohican or Mahican, Mohegan is actually the name of a distinct and separate East Algonquian–speaking Indian community in eastern Connecticut.

The name adorns a number of places in and around Greater New York, most notably Mohegan Lake, a pond and adjoining unincorporated village in the town of Yorktown, New York. William Jones, a local hotelier, first gave the name to what was then called Crompound Pond in 1859. The locale became a popular resort community during the late nineteenth and early twentieth centuries. An association of three hundred socially progressive families belonging to the Modern School Movement established their Mohegan Colony on the lake in 1930. Suburban subdivisions today occupy the former Mohegan Lake resort and colony properties. Mohegan Pond is located at the far eastern boundary of Greater New York in Fairfield.

MOHEPINOKE (Warren County, New Jersey). Mohepinoke Mountain is a 1,115-foot-high peak rising above the line of terminal moraine boulder rubble that covers the bedrock along the north side of the Pequest Creek in Liberty Township. The name Mohepinoke (also spelled Mohepinoki) Mountain first appeared with some regularity in geological reports assessing the mineralogical potential of the region published during the 1880s. The near uniformity of wording and spellings employed in sources subsequently mentioning Mount Mohepinoke, coupled with failure to find recognizable cognates elsewhere, suggest that a romantically inclined geologist or mining engineer may have introduced the name at this time. Further evidence of its synthetic origins comes from its construction from two Delaware words combined to produce the English translation "blood root." Whritenour points out that the actual Munsee word for the plant is *pekon*.

MOHICAN (Sullivan County, New York). Mohican is another very widespread Indian place name in the metropolitan area. It appears prominently as the name of the Mohican Lake and Catskills resort and residential community at the upper end of the town of Lumberland. Most nineteenth-century writers thought all Indians living on the east bank of the Hudson River spoke languages they interchangeably identified as Mohican and Mohegan. As mentioned earlier, Heckewelder (1876:52) wrote in 1818 that Munsees he lived with in the west called the Hudson River the *Mohicanichtuk*, their word for "River of the Mahicans." Publication of James Fenimore Cooper's hit novel *Last of the Mohicans* in 1826 caused the name to achieve a level of popularity that has hardly diminished over the intervening years.

MOMBASHA (Orange County, New York). Mombasha High Point is a 1,282-foot-tall peak in Sterling Forest State Park in the town of Tuxedo. Mombasha Lake, Creek, and village are located in the adjacent town of

Monroe. All are named after the Mount Bachon first recorded in George Clinton's 1735 field survey book for the Cheesecocks Patent (Freeland 1898:13). Mombasha is frequently regarded as an Indian name somehow connected with a Native namesake of Bashe or Bachom farther west in the Minisink Patent lands. Clinton, however, probably intended his name to be an allusion to the hills and country of Bashan where the biblical city of Golan stood along the present-day border separating Israel, Syria, and Lebanon.

MONHAGEN (Orange County, New York). Monhagen has been on maps of land in and around the city of Middletown at least since it was noted as the name of Monhagen Creek at the northeastern corner of the Wawayanda tract (French 1860:511). The name comes from County Monaghan in Ireland. Many residents who recognize the name's similarity to Mohican and are familiar with Monhegan Island in Maine tend to more closely associate it with Indians than with Ireland.

MONKA (Delaware County, New York). The 2,484-foot-high Monka Hill is a Catskill Mountain peak located at the northeast corner of the town of Middletown. Although often thought of as an Indian name, it is more probably the surname of a person of Scottish ancestry.

MONOMONOCK (Monroe County, Pennsylvania). Monomonock is a street name in the village of Mountainhome. The name was given to an inn operated in the village from the 1910s to the 1940s as a homage to the popular Monomonock Inn built atop the Watchung Ridge in Caldwell, New Jersey, that hosted catered affairs, conferences, and conventions between 1901 and 1942. Monomonock came to the region from New England, where lakes and other places in Massachusetts and New Hampshire still bear the name (in Bright 2004:295).

MONTOWAC (Morris County, New Jersey). Montowac Reservoir is located in Montville Township. It seems to be a local folk convergence conflating the Indian-associated Ojibwa word *Manitowoc* ("spirits") and the similar Polish word for "assembly or gathering" with the mountain reference in the first part of the township name.

MOOSEPAC (Morris County, New Jersey). Moosepac Pond is an artificial body of water located in Jefferson Township. It is probably a transplant from Moosabec Reach in Maine (in Bright 2004:297). Whritenour thinks *moosupeekw*, "elk pond," represents a Munsee equivalent of the Maine place name.

MOSHOLU (Bronx County, New York). Mosholu is best known as the name of a parkway and surrounding neighborhood in the Bronx. Mosholu Parkway is a broad, elegantly landscaped thoroughfare flanked by residential apartment buildings built during the early 1900s to connect the New York Bronx Botanical Gardens at Bedford Park with Van Cortlandt Park in North Kingsbridge. The name also survives as a widely used institutional and business name in the northwest Bronx and as Mosholu Avenue, an old road that today serves as a mixed residential and shopping street in the Riverdale section of the borough. Alert moviegoers will also remember the S.S. *Mosholu* as the towering four-master sailing ship featured prominently in the films *Rocky* and *The Godfather Part II*.

Many writers, including me (Grumet 1981:36), have regarded Mosholu as a local Indian word since Robert Bolton (1881[2]:446) first identified Tibbetts Brook as "the *Mosholu* of the Indians" in 1848. The Mosholu post office operated at the present intersection of Broadway and Mosholu Avenue throughout the middle decades of the nineteenth century. Although Indian, it is not a local name. It instead belonged to the prominent Choctaw leader Mushulatubbee (sometimes called Mashula) whose name combined the words *amoshuli*, "to persevere," and *ubi*, "to kill" (in Bright 2004:270). He was widely honored among Americans for his help during the War of 1812. The small village of Mashulaville in Noxubee County, located just a few miles from the Philadelphia, Mississippi, reservation of Choctaws who refused to remove to Indian territory, retains its name to the present day.

An abbreviated form of Mushulatubbee's name was given to the U.S.S. *Mashula*, a wooden screw steam sloop-of-war laid down in the New York Navy Yard for service in the Civil War in 1864. Somewhat indirectly, the S.S. *Mosholu* of movie fame also bears his name. Originally named the *Kurt*, she was built for German buyers in 1904 and confiscated while interned in an American port when the United States went to war with Germany in 1917. The memory of Mashula's perseverance probably lived on in the information given to First Lady Edith Wilson, who christened the ship supposed to be renamed Dreadnought with what she had been told was the Seneca Indian equivalent of the word "fearless." The ship itself still floats as a popular dockside restaurant moored along the Delaware River at Penn's Landing in Philadelphia.

MOUNT PAUL (Morris County, New Jersey). Local tradition holds that 815-foot-high Mount Paul in Chester Township at the southern end of Morris County is the burial place of the hill's Native namesake (Lenik 2009:110–11). Confirmation awaits discovery of records documenting some sort of association linking an Indian named Paul with the mountain. Another Mount

Paul, this one 1,210 feet high, is located at the northern end of the county at New Russia in Jefferson Township.

MUCKSHAW (Sussex County, New Jersey). Muckshaw Ponds Preserve is located in Fredon Township. Muckshaw is the surname of a family from the area and appears elsewhere as a nickname.

MUNSEY (Nassau County, New York). Munsey Park is an unincorporated community in the town of North Hempstead. It was named for its developer, Frank Munsey, who oversaw construction of the neighborhood during the 1920s. Various spellings of Munsey can be traced back to a widespread French Huguenot surname.

MUSKODY (Sullivan County, New York). Lake Muskody is an artificial pond located below Lake Tennanah on Trout Brook in the town of Fremont. The name comes from an Ojibwa word, *mashkode*, "prairie" (in Bright 2004:305). It is more widely known, however, for its appearance as Muskoda'sa, "the grouse," in Longfellow's poem *Song of Hiawatha*.

NAMEOKE (Queens County, New York). Trumbull (1881) suggested a Mohegan etymology, *name-auk*, "fishing place" or "where fish are taken," for Nameoke. Nameoke Street in Greater New York is located in Far Rockaway. Local businessmen using Indian names to attract clientele (a resort in the area was named the Tack-a-pou-sha Hotel) began selling lots in their newly named Nameoke Park development around 1905. Developers selected the name for its association with the then well-known Indian heroine Nameoke of the popular nineteenth-century romantic novel *Nix's Mate* (Dawes 1839). The name, connected as it was with Indians, romance, and fishing, caught on. By 1910 the local Democratic Party machine was calling itself the Nameoke Club after its meeting hall at the locale. Club members pointedly called the baseball team they sponsored the Indians. Memories of the associations that brought the name to prominence faded over time. It survives today as the name of a small side street lying in the shadow of a flight path at the southeastern end of nearby John F. Kennedy International Airport.

NAWAHUNTA (Orange County, New York). Lake Nawahunta in Harriman State Park was formed by enlarging Lemmons Lake on Lewis Brook in 1915. Park manager William A. Welch renamed the lake Nawahunta, from a Connecticut Mohegan word meaning "place of trout."

NEEPAULAKATING (Sussex County, New Jersey). Neepaulakating is a small creek that flows from Lake Neepaulin into Papakating Creek at the Sussex Airport at the hamlet of Lewisburg in Wantage Township. The conflated name was the winning selection in a contest held in 2004 by the Wallkill River Watershed Management Group and the Friends of Lake Neepaulin at the end of a two-year effort to select a name for the stream.

NEEPAULIN (Sussex County, New Jersey). Lake Neepaulin was built between 1955 and 1958 by a developer planning a lake club community at the locale. The name is a neologism that may identify the place as a née (i.e., former) or perhaps new Paulinskill. The lake is currently managed by an association of local homeowners known as the Friends of Lake Neepaulin organized in 1997.

NEPONSIT (Queens County, New York). Neponsit is a residential community located between Beach 142 and Beach 149 streets on the Rockaway Peninsula. Taking its name from the exclusive Massachusetts resort community of Neponset, the Neponsit Realty Company began selling lots at the locale in 1910. International attention briefly focused on the community when four U.S. Navy Curtiss seaplanes took off for the first transatlantic flight from its beach on May 8, 1919. Mechanical trouble forced three of the planes to withdraw from the effort. The single plane that made it splashed down in the English port of Plymouth nearly a month later, on May 31. Today, Neponsit is one of the very few New York City neighborhoods with an R-1 zoning rating that prohibits commercial structures of any kind and permits only single-family housing.

NIANQUE (Sullivan County, New York). Nianque Lake is located in the Ten Mile River Boy Scout Reservation. Originally called Wildcat Pond, the lake was created by damming a headwater of the East Branch of the Ten Mile River sometime before the Boy Scouts acquired and renamed the place in 1929. The name almost certainly comes from a word list identifying a Delaware word, *niankwe*, as "wildcat" or "bobcat."

NISKY (Northampton County, Pennsylvania). Often regarded as a truncated Indian place name, Nisky Hill Cemetery along the north shore of the Lehigh River just south of Moravian College in the city of Bethlehem more probably took its name from Niesky, a Bohemian village that was the hometown of many Moravian immigrants.

NOOTEEMING (Dutchess County, New York). Camp Nooteeming, the first Boy Scout camp in Dutchess County, moved twice from its original locale in the town of Fishkill before setting up at its current location in 1926 in Salt Point in the town of Pleasant Valley. Dillon Wallace, a writer of adventure stories and a founder of Scouting in Dutchess County who had made three wilderness excursions to Labrador, chose the Labrador Indian word he was told meant "men of the woods" as the suitably masculine name for a camp with the avowed purpose of turning boys into men.

OCQUITTUNK (Sussex County, New Jersey). Lake Ocquittunk is one of two ponds built and maintained by Civilian Conservation Corps workers in the present-day Stokes State Forest between 1933 and 1942. The lake was created by damming a section of the Big Flat Brook in 1934. Ocquittunk's origin is obscure. The name seems to most closely resemble the old Windham County, Connecticut, place name Ocquebituck, "top of a tree," in Trumbull (1881).

OHAYO (Ulster County, New York). The 1,388-foot-high Ohayo Mountain is in the town of Hurley; the nearby Ohayo Mountain Road runs from Hurley into the town of Woodstock. Local tradition holds that Ohioville, just south of New Paltz, got its name in memory of a would-be immigrant to Ohio who only got as far as Ulster County (Vasiliev 2004:165). It is not presently known if the Japanese-sounding spellings chosen for Ohayo and the nearby Tonshi mountains were deliberate or accidental.

OLEY (Luzerne County, Pennsylvania). Heckewelder (1834:360) thought Oley sounded like the Delaware words "*olink, wólink, olo,* or *wahlo,* 'a cavern cell, a sink hole; a dug hole to bury anything in, as also a tract of land encompassed by high hills.'" Five-mile-long Oley Creek is a tributary of the Nescopeck River at the very westernmost edge of the traditional Munsee homeland in Greater New York. The stream rises in Green Mountain four miles west of the Lehigh River village of White Haven and flows into Nescopeck Lake. The name at the edge of the Lehigh Valley is a comparatively recent transplant from Berks County, where Oley is the name of one of the earliest colonial settlements in the Upper Schuylkill Valley. Another example of a name from Berks County adorns Monocacy Creek in Northampton County.

ONAWA (Monroe County, Pennsylvania). Artificial Lake Onawa and several places near it are located in and around the village of Mountainhome in Barrett Township. The name comes from the mythical being Onaway in

Longfellow's poem *Song of Hiawatha* and is a popular place name adopted by local communities in several states.

ONTAROGA (Orange County, New York). The mid-nineteenth-century Ontaroga estate and present-day Club Ontaroga, a restaurant built in what were the estate's butlers' quarters, are located at the north end of the town of Goshen. The name was also given to Ontaroga Farm, a turn-of-the-twentieth-century summer retreat owned by a socially prominent Greenwich family in Ridgefield, Connecticut. Ontaroga is one of several Iroquoian-language place names from the Blue Ridge Mountains (others include Swannanoa and Watawga) evidently imported into the region to evoke romantic associations with Indians, picturesque scenery, and elegant country-resort living.

ONTEORA (Ulster and Greene counties, New York). The community of Onteora Park, Onteora Lake, Onteora Pond, Onteora Boy Scout Camp, and several other places in Catskill State Park currently bear the appellation that Schoolcraft claimed was the Mohawk name for the Catskills meaning "very high mountain" (in Bright 2004:353).

OPENAKA (Morris County, New Jersey). Openaka Lake is an artificial lake in Randolph Township. The Ninkey Forge, identified as the source for the name, was built by the dam at the mouth of the lake sometime during the mid-1700s. Although Openaka has often been identified as an Indian word, it is also thought to be an onomatopoetic representation of the sound made by the forge's power wheel. German shepherds bred at the Openaka Kennels, which opened by the old forge in 1923, provided many of the dogs trained as guide dogs by Dorothy Harrison Eustis at the Seeing Eye. This school, the oldest of its kind in America, opened in Morristown in 1929.

ORANGE (Essex County, New Jersey, and Orange County, New York). Several writers have suggested that Orange comes from Oringkes, the name of sachems of the Indians of Nevesincx noted in the March 28, 1651, deed to land at the mouth of the Raritan River (New Jersey Archives, Liber 1:7–8). The name instead honors the Dutch royal house of Orange that provided leaders for the Netherlands and one English king, who ruled as William III between 1689 and 1702.

ORONOQUE (Fairfield County, Connecticut). Oronoque is currently a community and street name in the town of Stratford. Placement of Oronoque and the Massachusetts name Woronoco on colonial maps were both probably inspired by the popular 1696 British play *Orinoko*, based on the widely read novel *Oroonoko* written in 1688.

OSCALETA (Fairfield County, Connecticut, and Westchester County, New York). Oscaleta Lake in the town of Lewisboro, New York, and Oscaleta Road, which runs across the state line from the lake into the town of Ridgefield, Connecticut, are often regarded as Indian names. Although Oscaleta sounds like Osceola, the name of the famous Seminole Indian leader that adorns several lakes and localities in the region, it is actually a Spanish word meaning "little kiss." The owner of what had earlier been known as South Pond renamed it after hearing the word while traveling in Spain.

OSCEOLA (Middlesex and Passaic counties, New Jersey; Westchester County, New York). Osceola, whose name meant "black drink shouter," was a much-admired Seminole war leader during the Second Seminole War. His notoriously deceitful capture in 1837 and his death from malaria a year later in an American prison touched the hearts of millions, who expressed their admiration by giving his name to places all over the country. Examples of such adoptions in Greater New York include street names in the towns of Wayne and Middlesex, New Jersey, and Osceola Lake, an artificial dammed body of water in the town of Yorktown, New York.

OSWEGO (Dutchess County, New York). Oswego Road runs along New York State Route 89 from the town of Union Vale west to the hamlet of Moore's Mill in the town of Lagrange. The name belongs to the important colonial-era harbor and trading post on Lake Ontario at the mouth of the Oswego River. Quakers brought it to Dutchess County to adorn their meetinghouse and cemetery (listed in the National Register of Historic Places in 1989) in present-day Moore's Mill in 1790. Residents continued to use the name for the community they established around the meetinghouse and as the name of the post office that operated at the locale until the mid-1800s (Kaiser 1965).

OUGHOUGHTON (Northampton County, Pennsylvania). Oughoughton Creek is a seven-mile-long stream that flows from the easternmost corner of Washington Township into the Delaware River in Lower Mount Bethel Township just south of Foul Rift. James and Linda Wright (1988:178–79) noted that the name first appeared at its present locale as O. Quartin's Creek on an early plat map of the area drafted in 1805 (on file in the Northampton County Hall of Records). The stream was noted as Richmond Creek in several documents penned between 1820 and 1840. Benjamin Miller (1939:69–70) wrote that the stream's name appeared as Oughquogton and Oquiston in otherwise uncited documents. The Wrights reported finding other records that spelled the name as Oquachton and Oquirton. Geological and hydrological reports dating from 1872 to the present day that mention Oughoughton under its current spelling limit themselves to technical assessments of issues impacting the creek and its watershed. It is difficult to assign an Indian identity to the name at present. The closest possible Munsee-language cognate seems to be Oghotakan, the name of a tract of land in the Wallkill River Valley in the state of New York near a waterfall Indians called Arachook. This tract was first mentioned in a June 4, 1699, mortgage pledge to land in the vicinity of the hamlet of Walden (New York State Library, Indorsed Land Papers [3]:60).

OWASSA (Sussex County, New Jersey). Owassa Lake, Dam, and Road are located just above the Bear Swamp Wildlife Management Area in Frankford Township. The name first appeared in a poem written in 1903 by a local clergyman recounting the legendary romance of Lenno and the love of his life, Owassa. The poem also transformed what was then prosaically called Long Lake into the idyllic Lake Owassa. Adopted as the lake's new name soon afterward, the name is most probably a more or less self-conscious adaptation of Owaissa, "the bluebird," from Longfellow's *Song of Hiawatha*.

OWEGO (Lackawanna, Pike, Susquehanna, and Wayne counties, Pennsylvania). The Milford-Owego Turnpike was chartered in 1821 to join the Delaware River community of Milford with the village of Owego on the North Branch of the Susquehanna River west of Binghamton. Named portions of the old turnpike still follow parts of the road's original route in the Greater New York area. These include the section called the Old Owego Turnpike in Milford and a much lengthier stretch known as the Owego Turnpike between Lake Wallenpaupack and Carbondale.

OWENO (Bergen County, New Jersey). Oweno Road and Place in Mahwah are all that remain of the Oweno estate named by its builder, wealthy industrial engineer and inventor Ezra Miller, in 1870. The name Oweno may be an Indian name from a place in Wisconsin where Miller worked in his youth as a surveyor. In 1884 Oweno replaced Mahwah as the local post office name for all of two months. One year later Miller died. His family kept the estate's great house going as an inn until 1899, when it burned down. Purchased by a developer in 1908, the Oweno estate was renamed Cragsmere, known among lovers of literature as the place where Joyce Kilmer wrote his poem "Trees." In 1938 village authorities ordered the property's owners to fill in badly polluted Oweno Lake, leaving the property's roadways as the only features preserving the name.

PECKMAN (Essex and Passaic counties, New Jersey). The Peckman River is a five-mile-long stream that runs from Verona in Essex County to its junction with the Passaic River at West Paterson. Often thought to be an anglicized version of a similar-sounding Delaware word, *pakim*, "cranberry," it is actually a family name belonging to local settlers.

PEEKAMOOSE (Ulster County, New York). Peekamoose Lake, Gorge, and other places bearing the name are located at the headwaters of Rondout Creek in Catskill State Park. The name is apparently a pun meaning something like "peek at or peak of moose." It was adopted by the Peekamoose Fishing Club established for exclusive private use of the locale in 1879. One of its founders, John Quincy Adams Ward (1830–1910), was a well-known sculptor whose fondness for romanticized Indian themes inspired his *Indian Hunter* (1868), the first sculpture created by a native-born American artist erected in New York's Central Park. The piece proved popular and was much copied. Peekamoose was part of a vernacular genre that produced at least one other joke name: Pok-o'-Moonshine, a boy's camp that opened on Long Pond near the western shore of Lake Champlain in 1905.

PEQUOT (Passaic County, New Jersey, and Nassau County, New York). Pequot appears as an Indian place name imported by modern-day developers to adorn ponds, lakes, roadways, and subdivisions throughout the United States. The name's value as an instantly recognized selling point rests on the Eastern Connecticut Indian nation's notoriety as the aggressive tribe that colonists defeated and until recently were widely thought to have been destroyed in the Pequot War of 1637–38. Examples in the metropolitan area include its use as a street name in Ringwood Township, New Jersey, and Port Washington on Long Island.

PHILLIPSBURG (Warren County, New Jersey). Residents of the freestanding town of Phillipsburg have maintained traditions that their community was named for an Indian chief called Phillip for at least 150 years. These traditions have been the basis for at least two origin stories. One holds that Phillipsburg was named for the Massachusetts Wampanoag sachem Metacomet, called King Philip by colonists, who led a coalition of Indian nations against New England colonists in 1675. The other links the community with a legendary Delaware chief named Phillip, said to have been the leader of an Indian community attacked by the very real Paxton Boys, rioters who indiscriminately assaulted and murdered Native people in eastern parts of Pennsylvania at the end of the last French and Indian War.

PISCATAWAY (Middlesex County, New Jersey). Piscataway Township in New Jersey is a transplant of Piscataqua, the place in New England that was home to many of the area's first colonists. New Jersey colonists frequently referred to the community as New Piscataway to distinguish it from other Piscataways in Maine and Maryland. Heckewelder thought that it was a Delaware word, *pisgattauwí*, "it is getting dark," that resembles similar words in other Eastern Algonquian languages.

PICATINNY (Morris County, New Jersey). Whritenour thinks Picatinny sounds much like a Munsee word, **pikahtunung*, "at the crumbling mountain." The name, however, appears to be a latecomer in an area fairly extensively documented by the beginning of the nineteenth century. It first appeared in the 1880 deed as the name of Lake Picatinny, the long-disused site of the Revolutionary War–era Middle Forge ironworks purchased by the U.S. Army for what became its Picatinny Powder Depot. Whatever its origins, today the name continues to grace Lake Picatinny, the 6,500-acre Picatinny Arsenal, and the Picatinny Peak that overlooks the lake.

PLUCKEMIN (Somerset County, New Jersey). Pluckemin has been the name of a hamlet in Bedminster Township since before the Revolutionary War. Often thought of as an Indian word, it more likely traces its origin to the village of Pluckemin in Scotland.

POMONOK (Queens County, New York). Pomonok is the name of a neighborhood and a public housing project located within it in South Flushing. The word Pommanocc first appeared as the nation of the sachem who conveyed title to Gardiner's Island to Lion Gardiner in 1639 (in Tooker 1911:182). The Montauketts and their neighbors used the name when talking about their homeland in present-day Suffolk County, New York. Another Suffolk County native, Walt Whitman, brought the name to international notice through his poem "Starting from Paumanok," which first appeared in his *Leaves of Grass* anthology edition published in 1860. For Whitman, Paumanok was a metaphor for the beginnings of America, his own start in life in the town of Huntington, and the beginning of his journey to poetically "strike up for a new world" (Whitman 1860). The name was later taken up by the members of the Pomonok Country Club in Queens County. Their greens have since become the center of the Pomonok neighborhood that is home to the City University of New York's Queens College and the Pomonok Houses, a public housing development built by the New York City Housing Authority in 1949. Pomonok is also a popular street and trail name in several communities; perhaps the best known is Suffolk County's 125-mile-long Paumanok Path.

POOTATUCK (Fairfield County, Connecticut). Like Mahican-speaking Catskills and Taconics who shared common borders with Munsee people, Pootatucks and their Wampano-speaking Pequonnock neighbors along the Connecticut–New York line intermarried with Munsees during colonial times. The Pootatuck State Forest in the Upper Housatonic River Valley is one of several locales in and around Greater New York preserving the name of this Native community.

PUNKHORN (Sussex County, New Jersey). Whritenour thinks that the punk in Punkhorn comes from an old English word for ash, powder, something that will smolder or ignite a substance—hence "powder horn." Today, Punkhorn Creek is a three-mile-long tributary of the Musconetcong River in Byram Township. Located where the Wright family began buying land during the late 1760s, Punkhorn started appearing with some frequency in late-nineteenth- and early-twentieth-century geological reports. The creek rises just to the south and east of Sparta and runs parallel to Sparta Mountain along Amity Road until it falls into Wright Pond. From there it flows into Wolf Lake and Lake Lackawanna, where it joins Lubbers Run on its way to the place where the run falls into the Musconetcong River just north of Waterloo Village.

QUANNACUT (Ulster County, New York). Quannacut Road and YMCA Camp are located in the hamlet of Ulsterville in the town of Wawarsing. The name was drawn from a list of Indian names from Long Island that Tooker (1911:301) thought were "suitable for country homes, hotels, clubs, motorboats, etc."

REVONAH (Sullivan County, New York). Revonah Lake, Hill (elevation 2,142 feet), and Resort are located in the town of Liberty. Constructed to power a sawmill around 1797, the lake was originally called Brodhead Pond. The resort's developer reversed the spelling of Hanover, his German birthplace, when he renamed the pond and the new tourist destination he built on its banks during the 1880s.

RICONDA (Passaic County, New Jersey). Sometimes identified as an Indian name and often spelled Rickonda, Riconda is the family name of the developer who built the present-day lake, subdivision, and service road in the unincorporated hamlet of Monks between the Monksville and Wanaque reservoirs.

RONDACK (Orange County, New York). Rondack Road in the city of Middletown bears the abbreviated version of the imported Mohawk name *atirun:taks*, "tree eaters." Noted in French colonial sources as Rondax, the name referred to Algonquin Indians who hunted and fished in what we now call the Adirondack Mountains.

SAGAMORE (Nassau County, New York). Sagamore is a New England Eastern Algonquian word for "leader." The name in Greater New York is most closely associated with the Sagamore Hill National Historic Site, established by Congress in 1962 to honor the memory of the locale's builder and most notable resident, Theodore Roosevelt. Roosevelt originally called the property Leeholm to honor his wife, who died while the estate was under construction in 1884. He subsequently renamed it Sagamore Hill after he remarried and moved in during 1887. The locale served as Roosevelt's summer White House between 1902 and 1908, and it remained his primary residence up to the time of his death in 1919.

SAGINAW (Sussex County, New Jersey). Lake Saginaw is an artificial body of water in Sparta Township. The name is an import from Saginaw, Michigan, derived from an Ojibwa word, *sa:ki:na:n*, "in the Sauk Country," from *osa:ki:*, "people of the outlet" (in Bright 2004:414–15).

SAMP MORTAR (Fairfield County, Connecticut). Samp Mortar Reservoir, Dam, and Brook are located in the town of Fairfield. Built in 1902, the reservoir and dam were named after the section of the brook running past rock ridges topped by small bowl-like depressions. Created by glacial scouring, these depressions have long been regarded as bowls that Indians used as mortars to grind corn into samp meal.

SAM'S POINT (Ulster County, New York). Sam's Point is the name of the spot where local resident Samuel Gonzalus is said to have made his life-saving leap to escape pursuing Indian raiders during the last French and Indian War. While Samuel Gonzalus was an actual resident in the area (Lake Maratanza was once called Gonzales Pond), Fried (2005:108–10) has shown that the story of Sam's escape was most likely cobbled together by a local mythmaker and published in the March 10, 1865, edition of the *New Paltz Times*. Today, Sam's Point continues to be a widely recognized Shawangunk Mountain landmark. The Nature Conservancy currently manages the 5,400-acre Sam's Point Preserve at the southern end of Minnewaska State Park.

SCHOCOPEE (Pike County, Pennsylvania). Schocopee Road runs past the one-room Schocopee Schoolhouse built in 1850 in Dingman Township. The name is probably an early import dating to the late 1840s from Minnesota, where the Dakota leader Shakopee welcomed missionaries to his community. Today, the chief's memory is most notably preserved as the name of the city of Shakopee on the Minnesota River southwest of Minneapolis.

SCIOTA (Monroe County, Pennsylvania). Sometimes regarded as a misspelling of the widely known place name Scotia, the name of the present-day village of Sciota is probably an import of Scioto, the name of a major river in Ohio frequently regarded as a Wyandot word for "deer."

SEBAGO (Rockland County, New York). Sebago Lake in Harriman State Park was created by damming Stony Brook Creek in 1925. The lake was named for the Sebago Lake resort community in Maine. The name comes from an Abenaki word, *masabagw*, "big lake" (in Bright 2004:427–28).

SENECA (Sussex County, New Jersey). Seneca Lake in Sparta Township is one of several occurrences of this name of the westernmost nation of the Iroquois confederacy imported into the metropolitan area.

SHACKAMAXON (Hunterdon and Union counties, New Jersey; Monroe County, Pennsylvania). Shackamaxon is an import from Philadelphia, where tradition holds that William Penn concluded a treaty with the Delaware Indians under an elm tree in 1682. Philadelphia today preserves the place where the tree fell in 1810 as a riverside public park. In Greater New York the name appears in Westfield, New Jersey, as the name of a lake, the golf and country club built around it, and a nearby street. It has also been adopted as a street name in the Clinton Township, New Jersey, hamlet of Annandale and in Pocono Lake, Pennsylvania.

SHAWNEE (Morris County, New Jersey; Nassau, Richmond, and Westchester counties, New York). Aside from its association with the historical Shawnee Indian community at the Delaware Water Gap between 1692 and

1728, Shawnee also serves as a transfer name for several places elsewhere in Greater New York. It appears most notably as the name of two lakes in places where Shawnees never lived: one in Yorktown, New York, and the other in Jefferson Township, New Jersey. Shawnee has also been imported for use as a street name in Staten Island, as the name of a drive in Massapequa, as a road name in Scarsdale, and as an avenue in Yonkers.

SHENANDOAH (Dutchess County, New York). Shenandoah is the name of a mountain, hamlet, and road in the town of East Fishkill. The hamlet of Shenandoah lies just below the north slope of Shenandoah Mountain near the junction of Interstate 84 and the Taconic State Parkway. Known earlier as Shenandoah Corners, a post office named Shenandoah served the locale between 1829 and 1856. The hamlet remains a residential community located at the southern end of the town of East Fishkill. Heckewelder reported that a Nanticoke chief told him that Shenandoah came from *schindhandowik*, a Delaware word meaning "the sprucy stream, a stream passing by spruce pines." Present-day scholars, however, mostly regard Shenandoah as a variously translated Iroquoian word. The name first appeared nearly simultaneously in upstate New York and Virginia. In 1712 New York settlers began purchasing land from Indians along Shenondohawah, the Mohawk name for the lower Mohawk River Valley above Albany (O'Callaghan 1864:118). That same year the Swiss colony builder Baron Christoph von Graffenried mentioned a mountain called Senantona while searching for a site for Palatine German colonists beyond Virginia's Blue Ridge in territory inhabited by Iroquoian-speaking ancestors of the Nottaways who presently refer to themselves as members of the Cheroenhaka Indian tribe (Todd and Goebel 1920:384). The word soon entered English as the name of the Shenandoah Valley. Residents of the Dutchess County hamlet of Shenandoah Corners were among many from other communities across the nation who adopted the name in hopes of capturing some of the pastoral allure associated with the valley in Virginia.

SHIPETAUKIN (Mercer County, New Jersey). Today, the five-mile-long Shipetaukin Creek flows through Lawrence Township from its headwaters at the north end of the village of Lawrenceville through the sixty-four-acre Shipetaukin Woods Preserve, managed since 1994 by the Lawrence Township Conservation Foundation, to its confluence with the Assunpink Creek just below Bakers Basin on the Delaware and Raritan Canal. The earliest occurrence of the name appears to be Shipetawkin (Gordon 1834:237). Its absence in the early records of a much-documented area, the possible association with a ship perhaps implied by the first part of the name, its location along the Delaware and Raritan Canal, and its appearance in 1834 just after the canal opened for business all suggest that Shipetaukin may well be a deliberately contrived "canal name."

SISCOWIT (Fairfield County, Connecticut, and Westchester County, New York). Siscowit Reservoir straddles the state line between New York and Connecticut. Most of the reservoir is in the town of Pound Ridge in Westchester; the dam at the reservoir's south end lies in Fairfield County. Placement of this name on regional maps as a replacement for the toponym Mead Pond following construction of Siscowit Dam in 1890 evidently relied on a fortuitous amalgam of distant and nearby names. The distant name was Siskiwit, a well-known mining and sportfishing center on the Upper Peninsula of Michigan (Bright 2004:447). Transfer of this name was eased by the reputation that the Mead Pond area enjoyed as a place of interest among geologists and anglers in Connecticut. Awareness of the fact that one of the Paugusset Indians who sold land a bit farther east in the town of Derby in New Haven County on March 10, 1710 (Philips 1901:441–43), was named Sissowecum may have helped ease Michigan's Siskiwit place name onto area maps in its currently modified form.

SKANANTATI (Orange County, New York). Skanantati Lake in Harriman State Park was created by damming an unnamed stream flowing from Lake Askoti into Lake Kanawauke. The name comes from a Mohawk word meaning "the other side."

SKENONTO (Orange County, New York). Skenonto Lake in Harriman State Park was formed by damming a stretch of Lake Sebago River in 1934. The word is a Mohawk name for the Hudson River.

SPRAIN (Westchester County, New York). Sprain Brook, Sprain Brook Parkway, Sprain Ridge Park, and Grassy Sprain Reservoir are only four of a number of Westchester locales bearing a name that many think is an Indian word. First appearing in Westchester in 1753, the name probably is the archaic English term *sprayne*, referring to the sowing of seeds by hand mentioned in several seventeenth-century deeds in Connecticut.

SQUANTZ (Fairfield County, Connecticut). Local tradition holds that the namesake of Squantz Pond and Squantz Pond State Park (opened in 1926) was a local chief who engaged in land negotiations with colonists in 1725 that ultimately led up to the April 24, 1729, purchase of land in the area. I have not been able to find a person identified by this name in documentation dating to 1725 that focuses on rumors of a possible local Indian attack (in Orcutt 1882:125–28) or in published sources documenting the 1729 purchase (De Forest 1851:360, which names only the three primary signatories). The name may come from Squantuck Swamp, mentioned in Naugatuck River Valley deeds farther east signed during the first decades of the 1700s (Orcutt 1882:42–43).

SQUAW (Passaic County, New Jersey). Squaw Brook, in North Haledon Township, is one of the dwindling number of places that still bears this word. Although the term is universally considered derogatory today, Goddard (1997) has shown that originally it was simply a New England Eastern Algonquian word for "woman," widely employed in the pidgin trade jargon spoken throughout the region.

STAHAHE (Orange County, New York). Stahahe Lake and Mountain are located in Harriman State Park. The lake was formed by enlarging Old Car Pond in 1918. One year later, park manager William A. Welch gave the name Stahahe to the pond, the brook that feeds into it, and the nearby mountain. The name comes from the Onondaga words *osdehee*, "rocks or ledge in water," or *osdehaehae*, "rocks on top" (in Bright 2004:461).

STARRUCCA (Susquehanna and Wayne counties, Pennsylvania). Trumbull's (1881) description of Scururra, a place name in Connecticut recorded in 1685 as an Indian name for Snake Hill (he proposed an etymology from a Delaware word *achgook*, "snake") suggests that Starrucca is another somewhat modified New England import brought into Pennsylvania by immigrants after the 1760s. Starrucca Creek flows mostly through Susquehanna County, New York, departing only to make a short bend into the far western end of Greater New York past the borough of Starrucca in adjoining Wayne County. Although it is widely regarded as an Indian word (cf. Donehoo 1928:21), its origin and affiliations are unclear. The name appears at its current locale as early as 1792 on Reading Howell's map of Pennsylvania. The Starrucca Viaduct is a stone arch bridge built between 1847 and 1848 to carry New York and Erie Railroad traffic across the Starrucca Creek valley near its confluence with the North Branch of the Susquehanna River at Lanesboro. The largest stone viaduct of its era and still in use, it was listed in the National Register of Historic Places in 1975 and is a Historic Civil Engineering Landmark.

SUSQUEHANNA (Warren County, New Jersey). Lake Susquehanna is an artificial lake in Blairstown Township. It was probably excavated to provide spoil to build up roadbed for the New York, Susquehanna, and Western Railroad, which had acquired the Blairstown Railroad in 1882 to support its line extending into the Susquehanna Valley coal fields. The Lake Susquehanna Airport, in operation since the 1940s, is now called the Blairstown Airport. Susquehanna is also a popular transplanted street and organization name in the metropolitan area. It first appeared in 1607 when John Smith's Powhatan guide identified Sasquesahanough as his people's name for the Susquehannock Indians they encountered at the head of Chesapeake Bay. Although Susquehanna is widely regarded as a Powhatan word meaning "muddy river," many linguists remain unsure of its etymology and affiliations.

SWANNANOA (Morris County, New Jersey). The former Petersburg Lake in the hills of Jefferson Township was acquired by circus impresario Alfred T. Ringling in 1900. A founder of the Ringling Brothers circus, which later became part of the merged Ringling Bros. and Barnum & Bailey Circus, Ringling purchased the property for his winter home. The developer who purchased the place six years after Ringling's death in 1919 renamed it Swannanoa in hopes of inviting flattering comparisons with Biltmore and other palatial country homes and resorts in North Carolina's Swannanoa Valley. The name itself probably comes from the Cherokee word *Suwali-Nunna*, "trail of the Suwali tribe," a reference to the Siouan-speaking Catawbas who called themselves *yi saaraa* (in Bright 2004:467).

TAHMORE (Fairfield County, Connecticut). Tahmore Drive is located in the Lake Hills development built in 1952 in the town of Fairfield. Furnished at the developer's request by the Fairfield Historical Society, it is the name of a legendary Indian princess, whose name was said to mean "hearts together" (Banks 1960:19–21). She was regarded as the daughter of chief Onee-to, a thinly disguised importation of German adventure novel writer Karl May's fictional character Winnetou, the Indian companion of the equally fictitious Old Shatterhand, into the region from across the Atlantic.

TE-ATA (Orange County, New York). Te-Ata Lake in Harriman State Park was created by damming a low-lying stretch of Popolopen Brook in 1927. Franklin and Eleanor Roosevelt named the lake to honor their friend, the well-known and highly regarded Oklahoma Chickasaw storyteller Te-Ata "Bearer of the Morning" (1895–1995).

TENNANAH (Sullivan County, New York). Long Pond at the head of Trout Brook was renamed Tennanah Lake in 1894 in hopes that the association with the well-known Alaskan Tanana River would attract sport anglers to the locale. The name is an Athabaskan word for "trail river" (in Bright 2004:478).

TICETONYK (Ulster County, New York). Ticetonyk Mountain (elevation 2,510 feet) was named for prominent local settler Mattys (nicknamed Tice) Ten Eyck.

TINICUM (Bucks County, Pennsylvania). The name of Tinicum Creek and Township in present-day Bucks County is another toponymic migrant into the region. Settlers presented an unsuccessful petition to incorporate a town identified by that name in Bucks County ten years before the 1737 Walking Purchase allowed proprietary authorities to claim title to the area; the petition was successfully resubmitted one year later. Tennicunk, the name proposed for the new township, was almost certainly drawn from well-known Tinicum Island much farther south, below Philadelphia in Chester County. A truncated version of Mattinacunck (Whritenour suggests this may be a Delaware word, *machtenachkunk*, "place of the bad fort"), the island had been the center of the New Sweden colony established exactly a century before Bucks County farmers made their first incorporation attempt much farther upriver (Battle 1887:193). The name in Bucks County is also preserved in Tinicum Township on several roads, organizations, and a covered bridge listed in the National Register of Historic Places in 1988.

TIORONDA (Dutchess County, New York). Many residents of Beacon think that the Tioronda Park and Street in their neighborhood take their names from what is popularly regarded as an Iroquois word for the nearby Fishkill Creek. It does appear to be an Iroquois name, adorning a falls on the Batten Kill more than one hundred miles farther north in Washington County, New York (Ruttenber 1906a:70). Joseph Howland, son of a wealthy New York merchant and soon-to-be Civil War general, introduced the name into the mid–Hudson Valley when he gave it to his newly purchased estate in 1859. It survives today as a street and park name in the city of Beacon and as the name of the one of the last bowstring truss bridges in the United States, opened in 1873 and listed in the National Register of Historic Places in 1976. The bridge's deteriorating trusses were dismantled and stored by the city of Beacon in 2006. Its stone abutments are preserved in place against the time resources are found to restore the bridge to its original position.

TIORATI (Orange County, New York). Tiorati Lake in Harriman State Park was formed by the damming of Tiorati Brook in 1915. The name comes from an Onondaga word, *dyowaeaede*, "where it is windy" (Bright 2004:496).

TOMAHAWK (Morris County, New Jersey, and Orange County, New York). Tomahawk is an Eastern Algonquian pidgin word for "hatchet or axe." The name most notably occurs in the metropolitan area as the Tomahawk Lake Water Park, first opened in New Jersey's Lake Hopatcong resort area in 1952, and as the Tomahawk Lake (built in 1929) and residential community in the town of Blooming Grove, New York.

TOMICO (Monroe County, Pennsylvania). Although Lake Tomico, located near the village of Canadensis in the heart of the Poconos resort country, bears the name of a prominent Munsee leader whose name adorns a road on the Moraviantown Reserve in Ontario, the place is probably named for a popular northern Ontario resort destination located above Lake Nipissing, and the name is probably from the Nipissing Algonquian language.

TONGORE (Ulster County, New York). Tongore Brook, Road, Park, and other places adorned by the name are located between the hamlet of Stone Ridge in the town of Marbletown and Olive Bridge in the town of Olive. Many local residents think Tongore is an Indian word. Like Tonshi below, Tongore may be a European place name built around the Old English words *tor*, "high, rocky hill," and *gore*, "triangular piece of land." The name first appeared in Spafford's gazetteer (1813:230) as the location of an iron furnace. The place was evidently named for the nearby Tongore Kill, a small rivulet today called Tongore Brook that flows through the village of Olive Bridge into the Esopus Creek. The Tongore Church and Cemetery were both established in 1823 at what was then called Olive. The Olive post office that first opened in 1832 briefly changed its name to Tongore in 1842 before taking up its present name of Olive Bridge in 1843. The name has since extended eastward to include Tongore Road and the Tongore Town Park just north of the road.

TONSHI (Ulster County, New York). Tonshi Mountain (more than 2,100 feet high) is located in the town of Olive near the Hurley and Woodstock town border. Nearby Little Tonshi Mountain (1,834 feet high) is wholly within Olive town bounds. The name was documented as Torneshook bergh (the first part of the name may translate as "hook-shaped hill"; its latter component is Dutch for "mountain") as early as the second quarter of the nineteenth century (in Evers 1988:183).

TOWAMENSING (Carbon County, Pennsylvania). Heckewelder thought Towamensing may have been a garbled version of *thuppekhanne*, a Delaware word meaning "a stream flowing from large springs, a spring issuing from springs in the earth." The present-day township name in Carbon County is almost certainly an import from the lower Delaware River just above Philadelphia. Towamensing initially appeared in William Penn's first Indian deed signed on July 15, 1682, as the name of a creek called Towsissinck at whose head stood the Indian town of Playwickey (Hazard 1852–60[1]:47–49). Repeatedly mentioned around Pennsbury Manor in lower Bucks County during the early eighteenth century, the name did not move north until 1759, when Nicholas Scull noted it as Toamensing at its current location on his map. William Scull noted Toamensong in the same place in his 1770 revision of his father's projection.

TUCCAMIRGAN (Hunterdon County, New Jersey). Tuccamirgan Park is a small municipal recreation area in the borough of Flemington. The park and a nearby road were named for Tuccamirgan, thought to be a Delaware chief who befriended Johann Philip Case, the first colonist to settle in Flemington. According to local tradition, he was buried in the Case family cemetery in 1750. In 1925 the Hunterdon Historical League honored his memory by raising a marble obelisk bearing his name in the cemetery. Its base is inscribed "In Memory of the Delaware Indian Chief Tuccamirgan 1750" on one side and "Erected by the Citizens of Flemington as a Tribute to this Friend of the White Man" on the other.

Sadly, like so many local traditions, this story is too good to be true. No records dating to colonial times mention a chief named Tuccamirgan (also said to be spelled Tuccaminjah, which Whritenour thinks sounds like a Northern Unami word, *ptukquiminschi*, "walnut tree"). The word is also claimed to be an otherwise undocumented Indian name for Mine Brook, which ran past the property purchased by Johann Case in 1738. The account of Tuccamirgan's burial, moreover, almost exactly matches a Besson family tradition describing an Indian funeral observed around 1750, when the Bessons settled in nearby Kingwood. This account was collected and published in the local newspaper by avocational historian John W. Lequear between 1869 and 1870 (reprinted in Moreau 1957:49–50). The wording of the Besson family tradition closely matched that found in earlier published accounts widely available in local libraries at the time. Other library sources capable of contributing to the legend included books like John Meginnes's *Otzinachson* (1889), a popular work that contained references to a Delaware Indian town identified as Tiquamingy, at the mouth of Chatham Run midway between present-day Jersey Shore and Lock Haven, where a local neighborhood intriguingly bears the name of Flemington.

TUCKAHOE (Westchester County, New York). Tuckahoe is the name of a village in the town of Eastchester. It is an old name in the area, given to the post office opened in the village in 1847 (Kaiser 1965). It is probably not the Chesapeake Bay Indian place name Heckewelder translated as *tucháchowe*, "deer are shy, difficult to come at," or *tuchausch-sóak*, "the place where the deer are very shy." No word resembling this place name occurs in regional colonial records. It is instead probably an import of another Chesapeake Bay Algonquian word, *tockwhogh*, widely used to identify an edible root often called Indian potato. The name also occurs elsewhere in the Northeast in southern New Jersey and Maryland.

TUSCARORA (Northampton County, Pennsylvania). Tuscarora Street in the city of Bethlehem is one of several locales bearing this name in and around the region. Driven from their North Carolina homeland by colonists after their defeat in the Tuscarora War of 1711–15, the Iroquoian-speaking Tuscaroras moved north into Pennsylvania, where they were admitted into the Iroquois confederacy as its sixth nation by 1722. Today, the Tuscarora Reservation in western New York near Niagara Falls serves as the nation's sovereign home territory.

UNAMI (Union County, New Jersey). Goddard (2010) identifies *wënáamiiw* as a Munsee word meaning "'speaker of Unami,' literally 'downriver person'; /wihwënaamíiwëw/ 'he or she speaks Unami'." As noted earlier, the Southern and Northern Unami dialects were spoken by people who mostly lived to the south of Greater New York during colonial times. This has not stopped many localities in the region from adopting the word as a transplant. Unami Park, a sprawling recreational facility in Cranford, Garwood, and Westfield townships completed by the Union County Park Commission in 1930, is among the more prominent places bearing this name in the region.

IMPORTS, INVENTIONS, INVOCATIONS, AND IMPOSTORS 239

UNCAS (Sullivan County, New York). Lake Uncas is located in the town of Rockland a few miles north of Livingston Manor in Catskill State Park. Camp Acadia for Catholic boys operated on the banks of the lake between 1907 and 1986. Today, it is part of the Monastery of our Lady in Beatitude owned by the nuns of the Order of Bethlehem. The historical Uncas was a Mohegan leader who played a major role in intercultural relations in eastern Connecticut during the seventeenth century. Places named Uncas outside of eastern Connecticut bear the name of the fictional son of the equally fictional Chingachgook, best known from the James Fenimore Cooper novel *The Last of the Mohicans*.

UNQUA (Nassau County, New York). Unqua Point, Lake, and Road in the hamlet of Massapequa Park in the town of Oyster Bay perpetuate a navigational error made on Long Island Sound that mistakenly noted Onkeway Point (in the present-day town of Fairfield, Connecticut) on the port rather than starboard beam of a New York–bound vessel. Unqua Point subsequently migrated directly south of that point to its present-day location alongside its more recently implanted local namesakes on Long Island's south shore.

UTSAYANTHA (Delaware and Schoharie counties, New York). Utsayantha Lake at the south end of the town of Jefferson in Schoharie County and nearby Mount Utsayantha (elevation 3,214 feet) in the Delaware County town of Stamford both lie at the headwaters of the West Branch of the Delaware River. Legend has it that the Mohawk princess Utsayantha and her son, born of a tragic liaison with an enemy of her people (either Indian or colonist, depending on who's telling the story), lie buried on the mountain.

WABASH (Passaic County, New Jersey). The Wabash Brook is a small tributary that flows through the Lake View development whose streets bear names of American states, cities, and rivers (in this case, Indiana's Wabash River). The small two-mile-long brook flows into the Passaic River a mile and a half below the Garden State Parkway Bridge that carries traffic across the river at Clifton.

WAKONDA (Rockland County, New York). Camp Wakonda is a summer camping facility in Harriman State Park for children operated by Homes for the Homeless. The camp has been at its present locale under a variety of managing organizations since the 1930s. A popular camp and resort name,

Wakonda comes from a Dakota word, *wakądą*, "to worship" (in Bright 2004:538).

WAMPUM (Monmouth County, New Jersey). Wampum Brook and Wampum Brook Park are located in the borough of Eatontown. The stream originally known as Mill Brook first surveyed by Thomas Eaton in 1670 was more recently given its present name to remind local residents of the cylindrical shell beads produced by Indians along the Jersey shore.

WANAKSINK (Rockland and Sullivan counties, New York). Wanaksink has been a place name in the Catskills at least since Ruttenber (1872:388–89) noted that the Reverend Charles Scott mentioned Wanoksink as a major Esopus settlement in a paper delivered to the Ulster County Historical Society. The origin of the name remains obscure. It was adopted in 1920 by the Wanaksink Lake Development Corporation, established to build the present-day lake and resort community mostly located in the present-day town of Thompson. The Lake Wanaksink Club presently manages the property as a private residential community. Similarly spelled Wanoksink Lake in Harriman State Park was created by damming the Christie Brook in 1934.

WANETA (Sullivan County, New York). Waneta Lake is an artificial lake located in the Willowemoc Wild Forest about 3.5 miles north of Livingston Manor. The name may commemorate a nineteenth-century Yanktonai Dakota leader who had been an opponent of American expansion before adopting a more conciliatory attitude during the War of 1812. It may also pay homage to the popular nineteenth-century song "Juanita."

WANGUM (Wayne County, Pennsylvania). Like many place names in northeastern Pennsylvania, Wangum Creek, Lake, Falls, and Falls Road in Palmyra Township are transplants from New England. The name probably comes from the southern New England Indian word *wangunk*, "a bend" (in Bright 2004:546). The name in Pennsylvania was likely directly borrowed from the Wangum Lake locale in Litchfield County, Connecticut.

WANOKA (Pike County, Pennsylvania). Wanoka Lake was named for the lake of the same name in Wisconsin. It comes from an Ojibwa word, *waanikaan*, "an excavated hole" (in Bright 2004:546).

WANTAGE (Sussex County, New York). Generations of New Jersey residents have regarded Wantage as an Indian place name since Heckewelder first proposed that it came from the Delaware words *wundachquí* or *undachquí*, meaning "that way." More probably, it comes from an English market town in Oxfordshire whose history dates back to Roman times.

WANTAGH (Nassau County, New York). Tooker (1911:273) notes that the residents of the Hempstead Township village of Ridgewood (originally called Jerusalem) renamed their community Wantagh in 1891 to honor the memory of Montaukett sachem Wyandanch, who involved himself in a land claim in the area during the 1650s.

WAPALANNE (Sussex County, New Jersey). Wapalanne is the name of a lake and camp built on its banks in Stokes State Forest. Both were constructed in 1933 by the Civilian Conservation Corps, who used the camp as its headquarters for its operations in the state forest until 1941. Wapalanne is a Delaware word for "bald eagle" almost certainly drawn from the entry for *woapalanne* in Brinton and Anthony's (1888:139) Lenape-English dictionary. Montclair State University opened its New Jersey School of Conservation at Camp Wapalanne in 1949 as its Environmental Field Education Campus. Today, Camp Wapalanne is the oldest continuously operating university-managed environmental education center in the nation.

WASIGAN (Warren County, New Jersey). Lake Wasigan and Camp Wasigan Road bear the name of the Freylinghuysen Township private co-ed Jewish summer camp operated at the locale until 1966. Identified as a Delaware word meaning "sunset" in a Bureau of American Ethnology (1926) list of popular Indian names compiled for summer camps, the name was selected by Camp Wasigan's founders in 1936.

WATABA (Fairfield County, Connecticut). Wataba Lake has been on Ridgefield town maps since local developers first gave the name to Rainbow Pond sometime around 1934. Although it sounds like a Mahican word, *wottapp*, "root," it was probably inspired by Watab Lake in Minnesota, translated there as an Ojibwa word specifically identifying tamarack or pine roots used for sewing bark canoes together (in Bright 2004:551).

WATAWAH (Monroe County, Pennsylvania). The glacial kettle hole known as Watawah Lake in the hamlet of Saylorsburg bears the name of the fictional mother of Uncas from James Fenimore Cooper's *Last of the Mohicans*. The name has been adopted elsewhere for a locale in Iowa and adorned the U.S. Navy Inspection Yacht *Watawah*, which was in commission during the first half of the twentieth century.

WATAWGA (Wayne County, Pennsylvania). Today, Watawga Lake is one of the cluster of artificially dammed ponds at the head of the Lehigh River in Lehigh Township. It is a slightly respelled version of Watauga, the Cherokee name of a Blue Ridge valley in eastern Tennessee where breakaway Virginia colonists set up their short-lived Watauga Republic in 1772. Developers imported the name in hopes of attracting buyers drawn by its romantic associations with Indians, mountain resort appeal, and rugged frontier independence. The Lake Watawga Association has managed the property as a private community since 1940.

WAUBEEKA (Fairfield County, Connecticut). Lake Waubeeka is located in the town of Bethel just a few miles south of the city of Danbury. The name does not appear in local colonial records and is evidently a late-nineteenth-century import from rural Wisconsin. Ridgefield town resident Hiram J. Kellogg probably named the lake when he built it to provide a height of water for a sawmill sometime around 1900. In 1950 the Naer Tormid ("Eternal Light") Society, an association of Jewish firefighters from New York City looking to start their own country community, purchased a 606-acre tract formerly owned by Kellogg that included the lake. Although they initially planned to rename the pond Lake Tormid, members instead kept the old name and gave it to their Lake Waubeeka Association. Waubeeka occurs elsewhere in New England, most notably as the name of the Waubeeka Golf links in Williamstown, Massachusetts.

WEONA (Northampton County, Pennsylvania). Weona Park and its well-known historic carousel (built in 1923) are located in the borough of Penn Argyl. It is sometimes identified as a form of Wyona, said to be an Indian word for elk. Local tradition holds that residents celebrated acquisition of the land on which the park was built by christening it "we own a." The Weona Park Carousel was listed in the National Register of Historic Places in 1999.

WETUMPKA (Somerset County, New Jersey). Wetumpka Falls in North Plainfield was named sometime during the nineteenth century for its considerably more imposing namesake in Alabama, which cascades from the lip of a scarp thrust up by a meteor strike eighty million years ago. The name comes from a Creek Indian word meaning "rumbling waters."

WE-WAH (Orange County, New York). We-Wah Lake was formed by two dams constructed between 1890 and 1895 on a small Ramapo River tributary to provide water for the village of Tuxedo in the town of Ramapo. We-Wah was the name of a Zuñi diplomat who represented his people at several meetings in Washington, D.C., around the turn of the twentieth century. A popular figure who traveled widely across the United States, he may have also been one of the first openly "two spirit" people (our closest equivalent is gay or homosexual) to effectively serve an Indian nation as a cultural ambassador.

WINNEBAGO (Morris County, New Jersey). The local Boy Scout Council in Elizabeth chose the name of a Siouan-speaking Indian nation from Wisconsin for the name of the camp they opened as the Winnebago Scout Reservation in 1950. Winnebago is an Algonquian name for a Siouan-speaking people. The word itself is thought to be Potawatomi: *winpyeko*, "person of dirty water," in reference to the periodically roiled state of a lake in the heart of their homeland. Members of the nation today prefer to call themselves Ho-Chunks, an anglicized version of their name in their own language variously translated as "big fish" or "big voice" (in Bright 2004:569).

WINNISOOK (Ulster County, New York). Winnisook Club and Lake in the Big Indian section of the town of Oliverea are located at the head of Esopus Creek in Catskill State Park. The Winnisook Club is one of the oldest private fishing clubs in New York. Members arranged for the construction of a log cabin (dubbed Winnisook Lodge) and a dam creating Lake Winnisook on club land in 1888. Situated 2,660 feet above sea level, Lake Winnisook is the highest lake in the Catskills. Club lore holds Winnisook was a seven-foot-tall Indian warrior chief who has since become the namesake of the nearby hamlet of Big Indian. True to the inexorably tragic trajectory of Victorian romance, club traditions still hold that his forbidden love for a local settler's daughter ended in his death and her inconsolable bereavement.

WINOCA (Fairfield County, Connecticut). Winoca Terrace is another of the Indian place names contributed in 1952 by the Fairfield Historical Society at the request of the owners of the Lake Hills development. Originally the name of an Ozark cave, translated in 1818 by Schoolcraft as an Osage word meaning "underground spirit," it also occurs as an early-twentieth-century streetcar suburb name combining the first two letters of the words constituting the name of nearby Wilmington, North Carolina.

WINONA (Morris and Sussex counties, New Jersey, and Orange County, New York). Winona Lake in Jefferson Township and Winona Parkway in Sparta Township are only two of several locales in New Jersey that bear the name. Another Winona Lake is located in the city of Newburgh in New York. These and other places bearing various spellings of the name honor Hiawatha's mother, a much beloved character in Longfellow's *Song of Hiawatha*. Winona also occurs in many other places across the nation and was a popular personal name for girls for many years, perhaps most widely recognized today as the name of the performers Winona Ryder and Wynonna Judd.

WISCASSET (Monroe County, Pennsylvania). The hamlet of Wiscasset, Wiscasset Road, and the Wiscasset Golf Course are all located just southeast of Mount Pocono in the resort town of Paradise. The name was imported from the picturesque Wiscasset community in the Boothbay region of Maine.

WISSAHICKEN (Passaic County, New Jersey). Heckewelder thought that the name of the Wissahickon Creek in Philadelphia, Pennsylvania, sounded a good deal like the Delaware words *wisamêkhan*, "catfish creek," and *wisaikhan/wisauchsícan*, "a stream of yellowish color." Wissahicken Brook is a small stream that flows into the Ramapo River in Wayne Township. The name is a slightly camouflaged import from Pennsylvania.

WYANDOTTE (Northampton County, Pennsylvania). Wyandotte Street (Pennsylvania State Route 378) in Bethlehem, Pennsylvania, was originally constructed in 1853 as a main thoroughfare for the Fountain Hill planned village, laid out with other streets bearing Indian names such as Seneca, Delaware, and Mohican. The steep hill scaled by the street has since become known as Wyandotte Hill (J. and L. Wright 1988:258–59).

WYCKOFF (Bergen County, New Jersey). Many Wyckoff Township residents believe that early spellings of the town name in forms such as Wikhoof and Wicough indicate that it was an Indian word (Wardell 2009:119). The place was neither the home of a family bearing the name nor a *wyckoff*, Dutch for "magistrate's town," during colonial times. The place ultimately grew large enough to support a post office of its own by 1826 and did finally rise to the dignity of a township in 1926. Despite the fact that it had no discernible local associations, Wyckoff is probably a Dutch place name introduced by settlers moving into the area during the early eighteenth century.

WYNOOSKA (Pike County, Pennsylvania). Wynooska Lake is located along the upper reaches of a tributary of Wallenpaupack Creek in Greene Township. Local tradition holds that the lake was built by Shaker families to power a tanning mill sometime during the 1800s. The death notice of the widow of Denis Olsommer (the couple were the progenitors of the family that ultimately bought the lake in the 1940s), which mentions her home at Wynooska Hill, represents the earliest known attestation to the name at the locale. Wynooska may be a slightly respelled reference recalling memories of Winooski, a Vermont locale well known to many of the New Englanders who first settled in the area. It may also be a somewhat respelled nod in the general direction of Minooka, the name of a neighborhood in the nearby city of Scranton.

WYOMING (Bergen and Essex counties, New Jersey). Wyoming Road in Paramus, Wyoming Avenue in Maywood, and Wyoming Station in Millburn are among the many place names in and beyond the borders of Greater New York that honor the state, pay homage to the beauty of Pennsylvania's Wyoming Valley, and preserve the memory of the violent battles fought in the Wyoming Valley during the Revolutionary War.

SOURCES

Manuscript Repositories

American Philosophical Society. Philadelphia, Pa.
Brooklyn Historical Society. Brooklyn, N.Y.
Connecticut State Archives. Hartford.
Fairfield Town Hall of Records. Fairfield, Conn.
Firestone Library, Princeton University. Princeton, N.J.
Goshen Public Library. Goshen, N.Y.
Library of Congress. Washington, D.C.
Monmouth County Hall of Records. Freehold, N.J.
Monmouth County Historical Association. Freehold, N.J.
Moravian Archives. Bethlehem, Pa.
Municipal Archives of the City of New York. New York.
New Jersey Archives. Trenton.
New Jersey Historical Society. Newark.
New Jersey State Library. Trenton.
New York State Library. Albany.
New-York Historical Society. New York.
Northampton County Hall of Records. Easton, Pa.
Orange County Historical Society. Arden, N.Y.
Special Collections, Alexander Library, Rutgers University. New Brunswick, N.J.
Sussex County Hall of Records. Newton, N.J.
Ulster County Hall of Records. Kingston, N.Y.
Westchester County Archives. Elmsford, N.Y.

Cited Publications

Allen, David Y.
    2006   "The So-Called 'Velasco Map': A Case of Forgery?" http://purl.oclc.org/coordinates/a5.htm.

Anonymous
    1872   *The Journall of the Procedures of the Governor and Councill of the Province of East New Jersey from and after the First Day of December Anno Dmni 1682.* Jersey City: John H. Lyon.
    1909   "Journal of New Netherland." In *Narratives of New Netherland, 1609–1664,* J. Franklin Jameson, ed., pp. 265–84. New York: Charles Scribner's Sons.
Backes, William J.
    1919   "General Zebulon M. Pike, Somerset-Born." *Somerset County Historical Quarterly,* Vol. 8, No. 4, pp. 241–51.
Bailey, James Montgomery
    1896   *The History of Danbury, Connecticut, 1684–1896.* New York: Burr Printing House.
Banks, Elizabeth V. H.
    1960   *This Is Fairfield, 1639–1640.* New Haven, Conn.: Privately published.
Barton, Benjamin Smith
    1798   *New Views of the Origin of the Tribes and Nations of America.* Revised 2nd ed. [First edition published in 1797].
Battle, J. H., ed.
    1887   *History of Bucks County, Pennsylvania.* Philadelphia: A. Warner and Company.
Beauchamp, William
    1907   *Aboriginal Place Names of New York.* New York State Museum, Bulletin 108. Albany: New York State Education Department.
Becker, Donald M.
    1964   *Indian Place Names in New Jersey.* Cedar Grove, N.J.: Phillips-Campbell Publishing Company.
Bischoff, Henry, and Mitchell Kahn
    1979   *From Pioneer Settlement to Suburb: A History of Mahwah, New Jersey, 1700–1976.* New York: A. S. Barnes and Company.
Bolton, Reginald Pelham
    1905   "The Amerindians of Manhattan Island." *Tenth Annual Report of the American Scenic and Historical Preservation Society,* Appendix C: 153–74.
Bolton, Robert
    1881   *The History of the Several Towns, Manors, and Patents of the County of Westchester, New York.* 2 vols. 2nd ed. New York [First edition published in 1848].
Boyd, Julian P., and Robert J. Taylor, eds.
    1930–71   *The Susquehannah Company Papers, 1753–1803.* 11 vols. Wilkes-Barre, Pa.: Wyoming Historical and Geological Society.
Boyd, Paul D.
    2005   "Settlers Along the Shores: Lenape Spatial Patterns in Coastal Monmouth County, 1600–1750." Unpublished doctoral dissertation, Department of Geography, Rutgers University, New Brunswick, N.J.
Bright, William
    2004   *Native American Placenames of the United States.* Norman: University of Oklahoma Press.

Brink, Benjamin Myer
　　1910　"The Patent of Rochester and the Settlement." *Olde Ulster,* Vol. 10, No. 4, pp. 97–104.
Brinton, Daniel Garrison, and Albert Seqapkind Anthony
　　1888　*A Lênâpe-English Dictionary.* Philadelphia: Historical Society of Pennsylvania.
Buck, William Joseph
　　1888　*William Penn in America.* Philadelphia: Friends Book Association.
Budd, Thomas
　　1685　*Good Order Established in Pennsylvania and New Jersey in America.* London.
Budke, George H., ed.
　　1975a　*Indian Deeds: 1630 to 1748.* New City, N.Y.: Library Association of Rockland County.
　　1975b　*Historical Miscellany: Volume 1.* New City, N.Y.: Library Association of Rockland County.
Bureau of American Ethnology
　　1926　*Circular of Information Regarding Indian Popular Names.* Washington, D.C.: Government Printing Office.
Campbell, Tony
　　1965　*New Light on the Jansson-Visscher Maps of New England.* The Map Collectors Circle, No. 24. London.
Christoph, Peter R., and Florence A. Christoph, eds.
　　1982　*Book of General Entries of the Colony of New York, 1664–1673: Orders, Warrants, Letters, Commissions, Passes, and Licenses Issued by Governors Richard Nicolls and Francis Lovelace.* Baltimore: Genealogical Publishing Company.
Christoph, Peter R., Florence A. Christoph, and Charles T. Gehring, eds.
　　1989–91　*The Andros Papers: Files of the Provincial Secretary of New York during the Administration of Governor Sir Edmund Andros, 1674–1680.* 3 vols. Syracuse, N.Y.: Syracuse University Press.
Cole, David
　　1884　*History of Rockland County, New York, with Biographical Sketches of Its Prominent Men.* New York: J. B. Beers and Company.
Cooper, James Fenimore
　　1826　*The Last of the Mohicans: A Narrative of 1757.* 2 vols. Philadelphia: H. C. Cary and I. Lea.
Cove, John J.
　　1985　*A Detailed Inventory of the Barbeau Northwest Coast Files.* National Museum of Man, Mercury Series, Canadian Centre for Folk Culture Studies, No. 54. National Museums of Canada, Ottawa, Ont.
Cox, John, Jr., ed.
　　1916–40　*Oyster Bay Town Records.* 8 vols. New York: Tobias A. Wright.
Crayon, Geoffrey [pen name of Washington Irving]
　　1839　"A Chronicle of Wolfert's Roost." *Knickerbocker,* Vol. 13, No. 4, pp. 318–28.

Davis, John, ed.
   1885   "The Trumbull Papers." *Collections of the Massachusetts Historical Society*, 5th Ser., Vol. 4, pp. 1–209.
Davis, William H.
   1876   *The History of Bucks County, Pennsylvania from the Discovery of the Delaware to the Present Time*. Doylestown, Pa.: Democrat Book and Job Office.
   1905   *The History of Bucks County, Pennsylvania from the Discovery of the Delaware to the Present Time*. 3 vols. New York: Lewis Publishing Company.
Dawes, Rufus
   1839   *Nix's Mate: An Historical Romance of America, Volume II*. New York: Samuel Colman.
De Forest, John W.
   1851   *History of the Indians of Connecticut from the Earliest Known Period to A.D. 1850*. Hartford, Conn.: Hammersley.
Donehoo, George P.
   1928   *A History of the Indian Villages and Place Names in Pennsylvania*. Harrisburg, Pa.: Telegraph Press.
Duncombe, Frances R., et al., eds.
   1961   *Katonah: The History of a New York Village and Its People*. Katonah, N.Y.: Historical Committee of the Katonah Village Improvement Society.
Dunn, Richard S., Mary Maples Dunn, Edwin B. Bronner, and David Fraser, eds.
   1981–86   *The Papers of William Penn*. 5 vols. Philadelphia: University of Pennsylvania Press.
Durie, Howard I.
   1970   *The Kakiat Patent in Bergen County, New Jersey*. Pearl River, N.Y.: Star Press.
Evers, Alf
   1982   *The Catskills: From Wilderness to Woodstock*. Rev. ed. Woodstock, N.Y.: Overlook Press.
   1988   *Woodstock: History of an American Town*. Woodstock, N.Y.: Overlook Press.
Folsom, Joseph Fulford, ed.
   1912   *Bloomfield Old and New*. Bloomfield, N.J.: Bloomfield Centennial Historical Commission.
Freeland, Daniel Niles
   1898   *Chronicles of Monroe in the Olden Time Town and Village, Orange County, New York*. New York: De Vine Press.
French, John Homer
   1860   *Historical and Statistical Gazetteer of New York State*. Syracuse, N.Y.: R. Pearsall Smith.
Fried, Marc B.
   1975   *The Early History of Kingston and Ulster County, New York*. Marbletown, N.Y.: Ulster County Historical Society.
   2005   *Shawangunk Place-Names. Indian, Dutch, and English Geographical Names of the Shawangunk Region: Their Origin, Interpretation, and Historical Evolution*. Gardiner, N.Y.: Privately published.

Furman, Gabriel
 1875 *Antiquities of Long Island*. Frank Moore, ed. New York: J. W. Bouton.
Gannett, Henry
 1894 *A Geographic Dictionary of New Jersey*. Bulletin of the United States Geological Survey, No. 118. Washington, D.C.: Government Printing Office.
Gehring, Charles T., ed.
 1980 *New York Historical Manuscripts: Dutch, Volumes GG, HH, and II Land Papers*. Baltimore: Genealogical Publishing Company.
 2003 *Correspondence of Petrus Stuyvesant, 1654–1658*. New Netherlands Documents Series, Vol. 12. Syracuse, N.Y.: Syracuse University Press.
Gibbon, Edward
 1851 *The History of the Decline and Fall of the Roman Empire*. William Youngman, ed. Paris, France: A. and W. Galignani [first published in 1776–89].
Gipson, Lawrence Henry
 1939 *Lewis Evans*. Philadelphia: Historical Society of Pennsylvania.
Goddard, R. H. Ives, III
 1997 "The True History of the Word Squaw." *News from Indian Country*, Mid-April, p. 19A.
 2010 "The Origin and Meaning of the Name Manhattan." *New York History*, Vol. 91, No. 4, pp. 277–93.
Golden, Jeff
 2009 "Identifying the Native People of the Red Hook, New York Area in 1609," pp. 1–19. jgolden@commonfire.org. Accessed July 3, 2012.
Goodrich, Phineas G.
 1880 *History of Wayne County*. Honesdale, Pa.: Haines and Beardsley.
Gordon, Thomas Francis
 1834 *Gazetteer of the State of New Jersey*. Trenton, N.J.: Daniel Fenton.
 1836 *Gazetteer of the State of New York: Comprehending Its Colonial History*. New York: Privately published.
Gould, Jay
 1856 *History of Delaware County and the Border Wars of New York*. Roxbury, N.Y.: Keeny and Gould.
Green, Frank Bertangue
 1886 *The History of Rockland County*. New York: A. S. Barnes and Company.
Greene, Don, and Noel Schutz
 2008 Shawnee Heritage: Shawnee Genealogy and Family History. 2nd ed. Lulu.com. Posted 2008. Accessed March 7, 2011.
Grumet, Robert Steven
 1981 *Native American Place Names in New York City*. New York: Museum of the City of New York.
 1988 "Taphow, the Forgotten 'Sakemau and Commander in Chief of all those Indians Inhabiting Northern New Jersey.'" *Bulletin of the New Jersey Archaeological Society*, No. 43, pp. 23–28.

1992 "The Nimhams of the Colonial Hudson Valley, 1667–1783." *Hudson Valley Regional Review*, Vol. 9, No. 2, pp. 80–99.

1993 "Magical Names." In *Unsettled Objects*, pp. 15–26 in a catalog published with the exhibition AMERICA Invention by Lothar Baumgarten, January 28–March 7, 1993. New York: Solomon R. Guggenheim Museum.

1994 "New Information from an Old Source: Notes on Adam the Indian's May 11, 1653 Testimony in the New Plymouth Colony's Records." *Bulletin of the New Jersey Archaeological Society*, No. 49, pp. 83–87.

1995 "The Indians of Fort Massapeag." *Long Island Historical Journal*, Vol. 8, No. 1, pp. 26–38.

1996 "Suscaneman and the Matinecock Lands, 1653–1703." In *Northeastern Indian Lives, 1632–1816*, Robert S. Grumet, ed., pp. 116–39. Amherst: University of Massachusetts Press.

2009 *The Munsee Indians: A History*. Norman: University of Oklahoma Press.

Handler, Richard, and Jocelyn Linnekin

1984 "Tradition, Genuine or Spurious?" *Journal of American Folklore*, Vol. 97, No. 385, pp. 273–90.

Harvey, Cornelius Burnham, ed.

1900 *Genealogical History of Hudson and Bergen Counties, New Jersey*. New York: New Jersey Genealogical Publishing Company.

Hasbrouck, Frank, ed.

1909 *The History of Dutchess County, New York*. Poughkeepsie: Samuel A. Matthieu.

Hazard, Samuel, ed.

1852–60 *Pennsylvania Archives Selected and Arranged from the Original Documents in the Office of the Secretary of the Commonwealth*. 1st series. 12 vols. Harrisburg.

Heckewelder, John Gottlieb Ernestus

1834 *Names Which the Lenni Lenape or Delaware Indians Gave to Rivers, Streams and Localities ... Within the States of Pennsylvania, New Jersey, Maryland, and Pennsylvania*. Revised by Peter S. Du Ponceau. Transactions of the American Philosophical Society, Vol. 4, pp. 351–96 [from a paper presented to the society in 1822].

1841 "Indian Tradition of the First Arrival of the Dutch at Manhattan Island, Now New York." *Collections of the New-York Historical Society*, 2nd Series, Vol. 1, pp. 68–74.

1876 *History, Manners, and Customs of the Indian Nations Who Once Inhabited Pennsylvania and the Neighbouring States*. 2nd ed. revised by William C. Reichel. Memoirs of the Historical Society of Pennsylvania, Vol. 12, Philadelphia [First edition published in 1819].

Hendrickson, Mark, Jon Inners, and Peter Osborne

2010 *So Many Brave Men: A History of the Battle of Minisink*. Easton, Pa.: Pienpack Publishing.

Hine, Charles Gilbert
    1908   *History and Legend, Fact, Fancy, and Romance of the Old Mine Road.* N.p.: Hine's Annual.

Historian, United States Postal Service
    2008   What's in a (Post office) Name? http://usps.com/postalhistory. Accessed April 6, 2009.

Hommedieu, L. E., ed.
    1915   *Proceedings to Determine Boundaries of the Wawayanda and Cheescocks Patents Held in 1785 at Yelverson's Barn.* Goshen, N.Y.: Independent Republican Printing Company.

Honeyman, A. Van Doren, ed.
    1927   *Northwestern New Jersey: A History of Somerset, Morris, Hunterdon, Warren, and Sussex Counties.* 4 vols. New York: Lewis Historical Publishing Company.

Hunter, Richard W., and Richard L. Porter
    1990   *Hopewell: A Historical Geography.* Titusville, N.J.: Township of Hopewell Historic Sites Committee.

Hunter, William A.
    1996   "Moses (Tunda) Tatamy, Delaware Indian Diplomat." In *Northeastern Indian Lives, 1632–1816,* Robert S. Grumet, ed., pp. 258–72. Amherst: University of Massachusetts Press.

Huntington, Elijah Baldwin
    1868   *History of Stamford, Connecticut.* Stamford, Conn.: Privately published.

Huntting, Isaac
    1897   *History of the Little Nine Partners of North East Precinct and Pine Plains, New York, Dutchess County.* Amenia, N.Y.: Charles Walsh and Company.

Hurd, D. Hamilton
    1881   *The History of Fairfield, Connecticut.* New York: J. W. Lewis.

Hutchinson, Viola L.
    1945   *The Origin of New Jersey Place Names.* Trenton: New Jersey Public Library Commission.

Jameson, J. Franklin, ed.
    1909   *Narratives of New Netherland, 1609–1664.* New York: Charles Scribner's Sons.

Jennings, Francis
    1975   *The Invasion of America: Indians, Colonization, and the Cant of Conquest.* Chapel Hill: University of North Carolina Press.

Johnson, Frederick C., ed.
    1904   "Diary of Brother John Martin Mack's and Christian Froelich's Journey to Wayomick and Hallobanck." In *Count Zinzendorf and the Moravian and Indian Occupancy of the Wyoming Valley (PA.), 1742–1763.* Reprinted from the *Proceedings and Collections of the Wyoming Historical and Geological Society,* Vol. 8, pp. 1–66.

Kaiser, Louis W.
    1965   *A Checklist of the Post Offices of New York State to 1850.* Ithaca, N.Y.: Privately published.

Kenny, Hamill
    1961   *The Origin and Meaning of the Indian Place Names of Maryland.* Baltimore: Waverly Press.
    1976   "Place Names and Dialects: Algonquian." *Names,* Vol. 24, No. 2, pp. 86–100.
Kipling, Rudyard
    1894   *The Jungle Book.* New York: Macmillan.
Kraft, Herbert C.
    1995   *The Dutch, the Indians, and the Quest for Copper: Pahaquarry and the Old Mine Road.* South Orange, N.J.: Seton Hall University Museum.
    2001   *The Lenape-Delaware Indian Heritage: 10,000 B.C. to A.D. 2000.* Elizabeth, N.J.: Lenape Books.
Kraft, Herbert C., and John T. Kraft
    1985   *The Indians of Lenapehoking.* South Orange, N.J.: Seton Hall University Museum.
Lambert, James F., and Henry J. Reinhard
    1914   *A History of Catasauqua in Lehigh County, Pennsylvania.* Allentown, Pa.: Searle and Dressler Company.
Leiby, Adrian C.
    1962   *The Revolutionary War in the Hackensack Valley.* New Brunswick, N.J.: Rutgers University Press.
Lenik, Edward J.
    2009   *Making Pictures in Stone: American Indian Rock Art in the Northeast.* Tuscaloosa: University of Alabama Press.
Lévi-Strauss, Claude
    1943   "The Art of the Northwest Coast at the American Museum of Natural History." *Gazette des Beaux-Arts,* Vol. 6.
Longfellow, Henry Wadsworth
    1855   *The Song of Hiawatha.* Boston: Ticknor and Fields.
MacReynolds, George
    1976   *Place Names in Bucks County, Pennsylvania.* 2nd ed. Doylestown, Pa.: Bucks County Historical Society.
Marshall, Donald W., et al., eds.
    1962–78   *Bedford Historical Records.* 4 vols. Bedford, N.Y.: Town of Bedford.
Masthay, Carl, ed.
    1991   *Schmick's Mahican Dictionary.* Memoirs of the American Philosophical Society, Vol. 197. Philadelphia.
Meginness, John F.
    1889   *Otzinachson: A History of the West Branch Valley of the Susquehanna.* Rev. ed. Williamsport, Pa.: Gazette and Bulletin Printing House [First published in 1856].
Miller, Benjamin L.
    1939   *Northampton County, Pennsylvania: Geology and Geography.* Pennsylvania Geological Survey, 4th series, Bulletin C 48. Harrisburg.

Mitchell, Joseph
    1949   "The Mohawks in High Steel." *New Yorker,* September 17, pp. 38–46.
Moreau, D. H., ed.
    1957   *Traditions of Hunterdon: Early History and Legends of Hunterdon County, New Jersey.* Flemington, N.J.: Privately published.
Morgan, Lewis Henry
    1851   *The League of the Ho-De-No-Sau-Nee, or Iroquois.* New York.
Myers, Albert Cook, ed.
    1970   *William Penn's Own Account of the Lenni Lenape or Delaware Indians: Tercentenary Edition.* Wallingford, Pa.: Middle Atlantic Press.
Nelson, William N.
    1902   "Indian Words, Personal Names, and Place-Names in New Jersey." *American Anthropologist,* Vol. 4, No. 1, pp. 183–92.
Nelson, William N., and Charles A. Shriner
    1920   *History of Paterson and Its Environs: The Silk City.* 2 vols. New York: Lewis Historical Publishing Company.
Nigro, Augustine
    2002   "Landscaping the Lehigh: The Creation of a Middle Industrial Landscape in Nineteenth-Century Pennsylvania." Doctoral dissertation, Department of History, West Virginia University. Morgantown.
Notorc
    2010a   What's in a Name? A Catalog of Indian Place Names in Westchester. http://notorc.blogspot.com. Posted May 7. Accessed April 8, 2011.
    2010b   Of Ponds and Lakes: A Catalog of Water Resources [in Westchester County]. http://notorc.blogspot.com. Posted May 12. Accessed April 8, 2011.
O'Callaghan, Edmund Burke, ed.
    1864   *Calendar of New York Colonial Manuscripts: Indorsed Land Papers in the Office of the Secretary of State of New York, 1643–1803.* Albany, N.Y.: Weed, Parsons, and Company.
O'Callaghan, Edmund Burke, and Berthold Fernow, eds.
    1853–87   *Documents Relative to the Colonial History of the State of New York.* 15 vols. Albany, N.Y.: Weed, Parsons, and Company.
Ogden, Lucile Gumaer, ed.
    1983   *The Journal of the Records of Peter E. Gumaer, 1771–1869.* Middletown, N.Y.: Privately published.
O'Meara, John
    1996   *Delaware-English/English-Delaware Dictionary.* Toronto: University of Toronto Press.
Orcutt, Samuel
    1882   *The Indians of the Housatonic and Naugatuck Valleys.* Hartford, Conn.: Case, Lockwood, and Brainard.
Palstits, Victor Hugo, ed.
    1910   *Minutes of the Executive Council of the Province of New York: Administration of Francis Lovelace, 1668–1673.* 2 vols. Albany, N.Y.: J. B. Lyon Company, State Printers.

Pelletreau, William S.
    1886   *History of Putnam County, New York.* Philadelphia: W. W. Preston.
Pennsylvania, State of
    1838–1935   *Pennsylvania Archives Selected and Arranged from the Original Documents in the Office of the Secretary of the Commonwealth.* 9 series, 138 vols. Harrisburg, Pa.
    1851–53   *Colonial Records of Pennsylvania.* 16 vols. Harrisburg, Pa.
Philips, Nancy O., ed.
    1901   *Town Records of Derby, Connecticut: 1655–1710.* Derby, Conn.: Sarah Riggs Humphrey Chapter, Daughters of the American Revolution.
Reading, John, Jr.
    1915   "The Journal of John Reading." *Proceedings of the New Jersey Historical Society,* Vol. 10, No. 1, pp. 34–46; No. 2, pp. 90–110; and No. 3, pp. 128–33.
Reed, Newton
    1875   *Early History of Amenia, New York.* Amenia, N.Y.: De Lacey and Wiley.
Reynolds, Helen Wilkinson
    1924   "Poughkeepsie, the Origin and Meaning of the Word." *Dutchess County Historical Society, Collections,* Vol. 1. Poughkeepsie.
Riker, James
    1852   *Annals of Newtown, Queens County, New York.* New York: D. Fanshaw.
Roome, William
    1897   *Early Days and Early Surveys of East Jersey.* 2nd ed. Morristown, N.J.: Privately published.
Rudes, Blair A.
    1997   "Resurrecting Wampano (Quiripi) from the Dead: Phonological Preliminaries." *Anthropological Linguistics,* Vol. 39, No. 1, pp. 1–59.
Ruttenber, Edward Manning
    1872   *History of the Indian Tribes of Hudson's River.* Albany, N.Y.: J. Munsell.
    1906a   "Footprints of the Red Men: Indian Geographical Names in the Valley of the Hudson's River ... Their Location and the Probable Meaning of Some of Them." *Proceedings of the New York State Historical Association,* Volume 6, pp. 1–241.
    1906b   "Cochecton." *Historical Papers of the Historical Society of Newburgh Bay and the Highlands,* Vol. 13, pp. 178–84.
Sahadi, Lou
    2011   *Affirmed: The Last Triple Crown Winner.* New York: Macmillan.
Salter, Edwin, and George C. Beekman
    1887   *Old Times in Old Monmouth.* Freehold, N.J.: N.p.
Sanders, Jack
    2009   Ridgefield Names: A History of a Connecticut Town Through Its Place Names. http://ridgefieldhistory.com. Posted January 10. Accessed May 9, 2010.
Sapir, Edward
    1949   "Culture, Genuine and Spurious." In *Selected Writings of Edward Sapir,* David Mandelbaum, ed., pp. 308–31. Berkeley: University of California Press.

Schenk, Elizabeth Hubbell
    1889   *The History of Fairfield County, Connecticut, from the Settlement of the Town in 1639 to 1818.* Fairfield, Conn.: Privately published.

Schoolcraft, Henry Rowe
    1845   *Report of the Aboriginal Names and Geographical Terminology of the State of New York, Part 1: Valley of the Hudson.* New York: New-York Historical Society.

Scott, Kenneth, and Charles E. Baker
    1953   "Renewals of Governor Nicolls's Treaty of 1665 with the Esopus Indians at Kingston, New York." *New-York Historical Society Quarterly,* Vol. 37, No. 3, pp. 251–72.

Selleck, Charles Melbourne
    1886   *Norwalk.* Norwalk, Conn.: Privately published.

Shea, Lisa
    1999   The Settlement of Philipsburgh. http://www.lisashea.com/genealogy/see/westhist.html. Posted January 18. Accessed November 11, 2011.

Siegel, Alan
    1989   "The Mystery of Dock Watch, or, What's in a Name?" *Warren History,* Vol. 1, No. 1, pp. 2–4.

Smith, J. Michael
    2000   "The Highland King Nimhammaw and the Native Proprietors of Land in Dutchess County, N.Y.: 1712–1765." *Hudson Valley Regional Review,* Vol. 17, No. 2, pp. 69–108.

Snell, James P.
    1881   *History of Sussex and Warren Counties, New Jersey.* Philadelphia: Everts and Peck.

Spafford, Horatio Gates
    1813   *A Gazetteer of the State of New York.* Albany, N.Y.: H. C. Southwick.

Speck, Frank G., and Jesse Moses
    1945   *The Celestial Bear Comes Down to Earth.* Reading Public Museum and Art Gallery, Scientific Publication No. 7. Reading, Pa.

Stillwell, John F., ed.
    1903–32   *Historical and Genealogical Miscellany: Data Relating to the Settlement and Settlers of New York and New Jersey.* 5 vols. New York: Privately published.

Stokes, Isaac Newton Phelps
    1915–28   *The Iconography of Manhattan Island, 1498–1909.* 6 vols. New York: Robert H. Dodd.

Street, Alfred B.
    1845   *The Poems of Alfred B. Street.* New York: Clark and Austin.

Street, Charles R., ed.
    1887–89   *Huntington Town Records.* 3 vols. Huntington, N.Y.

Strong, John A.
    1997   *The Algonquian Peoples of Long Island from Earliest Times to 1700.* Interlaken, N.Y.: Empire State Books.

Sullivan, James, et al., eds.
    1921–65 *The Papers of Sir William Johnson*. 15 vols. Albany: University of the State of New York.

Teller, Daniel Webster
    1878 *The History of Ridgefield: From Its First Settlement to the Present Time*. Danbury, Conn.: T. Donovan.

Todd, Vincent H., and Julius Goebel, eds.
    1920 *Christoph Von Graffenried's Account of the Founding of New Bern*. Raleigh, N.C.: Edwards and Broughton.

Tooker, William Wallace
    1896 *John Eliot's First Indian Teacher and Interpreter, Cockenoe-de-Long-Island, and the Story of His Career from the Early Records*. New York: Francis P. Harper.
    1901 *Indian Names of Manhattan*. New York: Francis P. Harper.
    1911 *The Indian Place-Names on Long Island and Islands Adjacent with Their Probable Significations*. New York: G. P. Putnam's Sons.

Trumbull, James Hammond
    1881 *Indian Names of Places etc. in and on the Borders of Connecticut, with Interpretations of Some of Them*. Hartford, Conn.: Case, Lockwood, and Brainard.

Van Valen, James M.
    1900 *History of Bergen County, New Jersey*. New York: New Jersey Printing and Engraving Company.

Vasiliev, Ren
    2004 *From Abbotts to Zurich: New York State Place Names*. Syracuse, N.Y.: Syracuse University Press.

Vermeule, Cornelius C.
    1925 "Some Early New Jersey Place-Names." *Proceedings of the New Jersey Historical Society*, Vol. 10, No. 2, pp. 241–56.

Wallace, Anthony F. C.
    1991 *Teedyuscung, King of the Delawares: 1700–1763*. Lincoln: University of Nebraska Press [First published in 1949].

Wampum, John, trans.
    1886 *Morning and Evening Prayer, the Administration of the Sacraments, and Certain Other Rites and Ceremonies of the Church of England; Together with Hymns (Munsee and English)*. Translated and Transcribed with the Assistance of H. C. Hogg. London: Society for Promoting Christian Knowledge.

Wardell, Patricia A., comp.
    2009 A Dictionary of Bergen County Place Names in Bergen County, New Jersey, and Vicinity. http://dutchdoorgenealogy.com. Accessed September 19, 2010.

Waterman, Kees-Jan, and J. Michael Smith, eds.
    2011 *Munsee Indian Trade in Ulster County, New York, 1712–1732: A Selection from the Anonymous "Account Book, 1711–1729"* [in Dutch], vol. 2 of *Philip John Schuyler Papers*. In press.

Whitehead, William A., et al., eds.
  1880–1931 *Documents Relating to the Colonial History of the State of New Jersey.* 2 Sers., 35 vols. Newark, Trenton, and Paterson, N.J.: Various printers.
Whritenour, Raymond
  1995 *Delaware-English Lexicon of Words and Phrases: Vocabulary.* Wayne, N.J.: Lenape Texts and Studies.
Wilk, Stephen R.
  1993 "Weequehela." *New Jersey History*, Vol. 111, Nos. 3–4, pp. 1–16.
Winfield, Charles H.
  1872 *History of the Land Titles in Hudson County, New Jersey.* New York: Wynkoop and Hallenbeck.
Wojciechowski, Franz L.
  1985 *The Paugusset Tribes: An Ethnohistorical Study of the Tribal Interrelationships of the Indians of the Lower Housatonic River Area.* Nijmegen, Netherlands: Catholic University of Nijmegen, Department of Cultural and Social Anthropology.
Woodland Indian Forum
  2010 Lenape Forum. http://woodlandindianforum.org. Accessed August 2, 2012.
Wright, George Edward
  1887 *A Visit to the States: A Reprint of Letters from the Special Correspondent of The Times.* London: The Times Office.
Wright, James, and Linda Wright
  1988 *Place Names of Northampton County, Pennsylvania.* Nazareth, Pa.: Privately published.
Wright, Kevin M.
  1994 *Map of Indigenous Place Names in Northeastern New Jersey.* River Edge, N.J.: Bergen County Historical Society.
Zeisberger, David
  1887 *Zeisberger's Indian Dictionary.* Eben Norton Horsford, ed. Cambridge, Mass.: John Wilson and Son.

www.ingramcontent.com/pod-product-compliance
Lightning Source LLC
Chambersburg PA
CBHW020832160426
43192CB00007B/620